The Trees that Made Britain

An Evergreen History

Archie Miles

1
BBC Books, an imprint of Ebury Publishing
20 Vauxhall Bridge Road, London SW1V 2SA

BBC Books is part of the Penguin Random House group of companies
whose addresses can be found at global.penguinrandomhouse.com

First published by BBC Books in 2006
This edition published in 2021

www.penguin.co.uk

A CIP catalogue record for this book is available from the British Library

ISBN 9781785946998

Printed and bound in Great Britain by Clays Ltd, Elcograf S.p.A

The authorised representative in the EEA is Penguin Random House
Ireland, Morrison Chambers, 32 Nassau Street, Dublin DO2 YH68

Penguin Random House is committed to a sustainable future for
our business, our readers and our planet. This book is made
from Forest Stewardship Council® certified paper.

For Reuben – may he come to know and cherish the trees in his world

Contents

Introduction

The treescape of the British Isles is a history lesson in its own right. Looking at the stories that these magnificent trees can tell us provides an incredible insight into many significant historical moments that have impacted on our present-day lives.

On 15 June 1215, King John signed the Magna Carta at Runnymede, but possibly at Ankerwycke on the east side of the River Thames. In 1532 it is believed that King Henry VIII dated Anne Boleyn under a specific tree at Ankerwycke. On 26 August 1346, King Edward III and his English army fought and won one of the most important battles in the Hundred Years War, near a village called Crécy in Northern France. Almost seventy years later, on 25 October 1415, King Henry V commanded between six and nine thousand troops and defeated the French army on a muddy battlefield at what we now know as the Battle of Agincourt.

What have these four events got to do with trees, you might well ask? The simple answer is that they are all related to one tree species: the English yew, *Taxus baccata*, one of only three conifers native to the British Isles.

The Ankerwycke yew is an iconic 2,500-year-old ancient yew tree growing by the River Thames, and the two battles in France were won by the formidable English bowmen who defeated the French knights used powerful longbows made from a single piece of timber, cut from the English or common yew. Despite a majority of the raw bow staves coming from Europe, without the yew tree, the English army would almost certainly have been defeated at both Crécy and Agincourt, and Europe would have been a very different continent today.

The average draw-weight of a yew longbow is around 120 pounds. I was once given the opportunity to shoot an arrow from a traditionally-made yew longbow by Robert Hardy, an authority on medieval warfare and longbows, but I struggled to draw a 40-pound bow. Since then I have so much respect and admiration for all the bowmen that fought for us on the French battle fields.

Built in 1512, the *Mary Rose* was Henry VIII's naval flagship for 34 years, a carrack-type warship built in Portsmouth using a huge amount of English oak timber. Some of this was cut from oak trees in the neighbouring Bere Forest in Hampshire, but much also came from the Wealden Forest to the north, as mature English oaks, *Quercus robur,* were already becoming a rarity around the shipyards due to large scale felling for ship building and building construction. It is estimated that over 600 mature oak trees covering a 16-hectare forest would have been felled to build this incredible warship, leaving a huge void of trees in our ancient woodland. Unfortunately, the *Mary Rose* sank in 1545 fighting the French during the Battle of the Solent with the entire crew of around 500 on board, She remained on the bed of the Solent for 437 years until raised in 1982 and preserved in a museum in Portsmouth. After having the privilege of seeing the natural form of the tree's branches used in the construction of this ship at close quarters in the *Mary Rose* Museum recently, I now walk through oak woodlands wearing the hat of the sixteenth-century forester who would have been sent out with axe and saw to bring back each individual piece for the ship builders to construct the huge frame.

My favourite tipple is a dram of scotch whisky, and I soon learnt that it cannot be called whisky until the raw spirit has been matured in a cask made from oak for at least three years. However, the longer it's in there, the better the taste and flavour of the whisky. Oak is the preferred timber due to its strength, durability, and liquid-tightness, making it the favoured wood for coopering into a whisky barrel.

At a well-known whisky distillery in Scotland I was given a hands-on lesson by the master-taster in 'toasting' – charring the inside of the barrel to release the vanillin and tannins from the wood that help to turn the spirit red during whisky maturation. Now when I sip my wee dram, I can taste the oak, like the smell of sawdust from a light sanding of a solid oak tabletop, and I appreciate this iconic British woodland tree even more than I ever did before.

Ash, *Fraxinus excelsior*, is another iconic tree on the British landscape and probably has had more uses economically in our society than any other tree, helping to change and advance the course of our history. Its wood is not as durable as oak, but it is a light creamy-white to pale brown colour, and has the advantage of a flexible and shock-resistant wood grain. This makes it the ideal material for tool handles such as garden spades and forks, the woodman's axe and carpenter's hammer, and for sports equipment such as tennis rackets, hockey sticks, Irish hurleys and snooker cues.

It was also an important construction material in the early aviation and motor car industries, as seen in the de Havilland "Gypsy Moth" aeroplanes and the Morgan car. In fact ash has been used in almost every sector of the manufacturing world, including chair-making, where the different parts of the chair such as the spindles were turned on a traditional pole lathe by the bodgers of the woodlands.

This amazing tree replaced the lost elm as the primary hedgerow tree in the British countryside following Dutch elm disease in the nineteen sixties and seventies but unfortunately is now itself under serious threat from an introduced fungal disease called ash dieback (*Hymenoscyphus fraxineus*). Since arriving on our shores in 2012, dieback has moved across the British Isles rapidly, seriously affecting both young and mature ash trees, potentially killing 90–95% of the ash population in the coming years and threatening many species of biodiversity that rely on this tree as a host.

Waiting in the wings, although not yet on our shores, is a further threat to the ash: an exotic beetle borer named the emerald ash borer (*Agrilus planipennis*). Let us hope that this beetle never reaches us and that there will still be healthy living ash trees left in our hedgerows and woodlands for the next generations of our families to be able to see and use.

I have only mentioned three native British tree species in this introduction and there are so many more with equally interesting historical facts and stories to tell. I hope that this beautifully illustrated book written by my very good arboreal friend, Archie Miles, will allow every reader to realise the significant impact that trees have had on all of our lives and how these trees have made Britain.

Tony Kirkham MBE, VMH, AoH

Oak

The quintessential tree of Britain is the oak, symbol of the most enduring and admirable facets of the nation's rich cultural heritage and landscape history. The oak tree is synonymous with strength, resolution, dependability and endurance, and what we as a nation see in our national tree is arguably what we would expect of ourselves and others.

Sing for the oak-tree,
The monarch of the wood,
Sing for the oak-tree,
That groweth green and good;
That groweth broad and branching
Within the forest shade,
That groweth now, and yet shall grow,
When we are lowly laid!

FROM 'WHEN WE ARE LOWLY LAID'
BY MARY HOWITT (1799–1888)

Acorns of the sessile oak.

Steeped in history, the references to oak come thick and fast. The Boscobel Oak, where King Charles II took refuge, commemorated to this day by an abundance of Royal Oak public houses. English diarist John Evelyn's 'wooden walls', for which he entreated the nation to plant oaks in his *Sylva* of 1664. The great actor-manager and playright David Garrick's patriotic lines – 'Heart of oak are our ships, heart of oak are our men' – addressed to a navy who were those 'wooden walls' in defence of Britain. To the victor and the valiant the wreath of oak leaves (harking back to Roman times). Oak leaves adorning medals for bravery and coins of the realm. Great oaks from little acorns grow – the acorn, the ultimate, natural powerhouse, symbol of industry, achievement and fecundity. True to form, English herbalist Nicholas Culpeper excused himself any need to describe the tree, 'It [the oak] is so well known (the timber thereof being the glory and safety of this nation by sea) that it needs no description.'

TWO OAKS

Britain actually has two native species of oak – the English oak (*Quercus robur*) and the sessile or durmast oak (*Quercus petraea*) – though they were not formally differentiated until the early nineteenth century. The English oak, often called the common oak or pedunculate oak, has a vast native range, stretching throughout

Europe and Asia Minor, from Scandinavia to the Caucasus. It will thrive on most soils as long as the conditions are not too acidic or waterlogged and in Britain it is predominantly a lowland species. It is a broad-spreading tree and its two most distinctive features are the leaves, which have very short stems, and the acorns, borne on long stems or peduncles. This is the oak that traditionally produced the best timber for shipbuilding. It also played an important role in the ancient system of pannage – the legal right to send one's animals, especially pigs, to forage in the woods. Acorns appear in abundance after the age of ten or so and this is relatively early compared to the sessile oak, which takes about 40 years to produce its first crop.

The sessile or durmast oak does best in sheltered situations and it grows well on damp, upland sites. It is generally a more compact, upright tree than the English oak but in exposed locations, such as along sea cliffs or on mountainsides, its natural shape will become distorted. Its leaves are borne on short stems and they have a downy underside that appears silvery or creamy white in the light. The acorns are sessile, that is they have no stems at all, which is the feature that gives the tree its common name (from the Latin *sessilis* meaning 'low of sitting'). Its native range is the same as for the English oak, although it is thought to be the only oak native to Ireland. While it is generally accepted that true hybrids of the two species are seldom found, they will often produce a wide variety of intermediates in places where both are growing in close proximity.

Oaks have a remarkable capacity to throw a second flush of foliage in midsummer, usually to replace leaves lost to moth-larvae infestations. This is known as Lammas growth, so called because it usually occurs around 1 August (Lammas Day) when the first harvesting of corn was traditionally baked into bread and

An English oak in isolated splendour.

consecrated. During this time the trees will bear two distinct sets of leaves, the older foliage having matured to a dark green, contrasting with the bright green (or in some cases slightly reddish) colour of the new.

LIFE AND DEATH

Oaks are good colonizers, particularly in open ground, though young seedlings tend not to fare so well within woodlands because of the deep shade and the damage caused by small mammals. It is well recorded that the oak's most helpful agent of colonization is the jay, for in the autumn these birds will busy themselves collecting acorns, which they will then bury for their winter larders. It has been observed that a single jay is capable of gathering and storing up to 5000 acorns during one ten-week period. Equally amazing is their ability to find the acorns once again during the winter months, often beneath many inches of snow. However, the birds don't recover them all, as mice will help themselves, some will rot, and others will be left behind to germinate.

It is estimated that around 500 species of invertebrates are reliant on the oak. Add to that the birds and bats that nest and roost in the trees and feed on attendant invertebrates, plus the intimate, mycorrhizal relationships with various fungi, along with purchase for vast arrays of mosses, lichens and ferns, then it becomes clear that the oak is of pivotal importance in many habitats. There is a greater variety of galls – excrescence produced by insect, fungus or bacterium – to be found on English oak than any other single tree or plant species in Britain. Most are caused by a minute invasion into the tree's tissue by small gall-wasps during the process of laying their eggs. This then stimulates the tree to produce mutated tissue that creates the specific forms of galls for each individual wasp

A stag-headed oak in the medieval deer park at Moccas in Herefordshire.

species with the protective and nutritional requirements of their larvae. These tiny galls in many different shapes and sizes have an intrinsic beauty all of their own. Most familiar are the knobbly globes of oak apples and the smoother, and slightly smaller, marble galls, both of which form on the buds. Look closer and find perfect little discs of various spangle galls beneath the leaves, currant galls on flowers and the small hairy tufts of artichoke galls on buds. The knopper galls attack acorns and this may affect their reproductive viability where they occur in large numbers, but most galls do no lasting harm to the host tree.

It is quite common to see older oak trees where the canopy has receded, leaving dead branches poking out beyond the live foliage. This 'stag-headed' appearance is unsightly but it doesn't necessarily mean that the whole tree is going to die. Oak has a remarkable capacity to retrench in old age and this is its way of reducing the demands on its root system. Most trees will continue to put on small amounts of new growth, but within the confines of a more compact canopy.

OAK AND LIME IN THE LANDSCAPE

The popular perception of our oakwoods today is that they are the remnants of a historic, national forest that stretched blanket-like across most of lowland Britain for thousands of years until farmers decided to start making clearings in which to grow crops, keep livestock and build settlements. About 9500 years ago, as many colonizing trees were still finding the extremities of their native ranges, it's more likely that the vegetation was a mixture of trees and grassland (savannah). By this time the Mesolithic hunter–gatherers would have started to have some effect on the existing tree cover by cutting wood for fuel and to make shelters.

The Cambridge botanist Oliver Rackham claimed that human beings have been the principal agents of change from the very beginning, although clearly aided by the needs of their domestic livestock. The Dutch systems ecologist Frans Vera, on the other hand, believes that European savannah was shaped by large, wild herbivores in a complex pattern of grazing and woodland regeneration, with the oak playing a central role as a pioneering species, aided by the protection of nurse trees such as hawthorn and blackthorn. There is credence to both theories but Vera's findings are very much based on a European model, which includes different species of tree as well as different herbivores from those that would have been found in Britain. The debate will continue. During the period between 5000 and 7000 years ago, when Britain was experiencing its 'climatic optimum' (the so-called 'Atlantic' period, when mean temperatures were about 4.5 ° F warmer than today),

These small-leaved lime coppice stools in Bedford Purlieus, on the Northamptonshire and Cambridgeshire border, may well be several hundred years old. Part of Rockingham Forest, the name purlieu signifies an annexed or detached part of the forest.

British woodland was at its peak, not only with the oak, which by this time had made it all the way to Scotland, but also with ash (*Fraxinus excelsior*) and the small-leaved lime (*Tilia cordata*).

During this time the small-leaved lime was at the northern limit of its range, yet the climate still allowed it to reproduce from seed. Pollen records indicate that it was probably the dominant woodland species across much of lowland Britain. As temperatures slowly cooled over thousands of years this tree, and its cousin the large-leaved lime (*Tilia platyphyllos*), ceased to produce viable seed and, as humans slowly tamed the wild wood for their own requirements, oak and ash outstripped limes as the universally most useful timbers and so little effort was made to plant it. The sparse and localized populations of the native limes that remain today will occasionally reproduce from layering boughs and sometimes they will regenerate from windthrown trees bedding into the woodland floor. More commonly, many of them are simply survivors from the continual intervention of coppicing regimes. There are some massive lime coppice stools in Britain that measure over 50 feet across, the single stools, the root or stump of a felled tree, having spread and fragmented to form circles that look as though they are individual trees. Some of these may go back several thousand years, vying with the yews for the greatest antiquity. A strange and, in this case, not altogether unwelcome consequence of recent climate warming in Britain has been that many native limes in southern Britain are once again beginning to set viable seed. Could this signal the resurgence of this species?

In many respects, up until about 1850, much of the New Forest in the south of England was very like the savannah system, which held sway because of good forest management. Open areas of grassland, wood-pasture and stands of dense woodland combined to make a balanced mosaic, a sort of parkland that worked for the commoners and their livestock, the king and his huntsmen and for the wildlife. Once dominated by oak, the last two centuries have seen the 3:1 ratio in its favour over beech eroded to equal parity. In much of the dense, woodland shade beech will thrive in preference to oak, thus changing the character of these woods. Even so, there are still many delightful pockets of woodland in the New Forest that have the typical oak and beech mix with an understorey of holly, which has always made an ideal nurse for the seedlings. There have been serious issues in the past concerning overgrazing by deer and ponies, the massive damage that grey squirrels can cause and the pressures of increased visitor numbers, but these will now be tackled because the New Forest acquired national-park status in 2005 and now there is a new management regime in place with a committed, published strategy for the future.

Oak exists as a constituent species in many different types of woodland and it is fair to say that it has always been managed, either by deliberate planting, coppicing, pollarding or felling. Plantation oaks are usually evenly spaced, often after selective thinnings, and they are grown quite close together to discourage branching. They will all be harvested before they have seen a century. Coppiced oaks are, in most cases, a vestige of woodland management that had become largely irrelevant by the beginning of the twentieth century. Up to that time vast amounts of oakwood were needed to make the charcoal to fire the furnaces to make iron. Coke, the solid substance left after heating coal, or petrol, made these uses redundant. Coppiced oakwoods used to be more useful for their tan bark than their timber but synthetic tanning agents put paid to this industry. Fuel wood for the home was eclipsed by electricity, gas and oil. Thankfully, a few coppiced oakwoods are still worked, mainly for specialized and small-scale industries, such as furniture-making, wood-turning and cleft-wood gates and fences.

Constructing the stack for a traditional charcoal burn in the Forest of Dean. An air-tight layer of turf covers the wood to be turned into charcoal. Since this is for a demonstration this stack is much smaller than the working originals would have been.

Charcoal-making has returned in some parts of the country (alder is also used for this) and there is a tannery in Colyton in Devon that uses tan bark.

Pollarded oaks are usually large, old trees, either in woodlands or in open parkland, where they have often been part of old wood-pasture regimes. They survive on boundaries and in hedgerows as former marker trees and some of them have been left as grand old relics in the middle of a field, farm or village. Unlike maiden trees, which do well to make 300 or 400 years, some ancient pollards are twice this age and a few are well over 1000 years old. They survive purely because they have been regularly cut back over the centuries and so they have a relatively compact structure to support.

Some of the most spectacular oakwoods are to be found along the west side of Britain. Often known as the 'Atlantics', these woods are largely of sessile oak and they usually occur in rugged upland sites, where sheltered hollows and permanently damp and frost-free gorges create an ideal microclimate. These conditions are also good for mosses, liverworts and ferns. On top of Dartmoor there are three ancient woods that have all these characteristics: Wistman's Wood, Piles Copse and Black Tor Copse. Unusually, English oak is the predominant species here rather than sessile oak but because the woods are so old no one knows why this is the case, especially as there are sessile oakwoods all around Dartmoor's perimeter. One possible explanation is that English oaks might have been introduced deliberately for their superior acorn crops. The trees in these three woods are thought to be more than 1000 years old. They are smaller than might be expected, indicating that they would have been coppiced in the past; the extreme weather conditions would also have helped to slow down their growth. Centuries of biting gales, snow and ice have shaped these trees into stunted, twisted corkscrews dripping with mosses and lichen but their rocky domain has kept them safe from grazing sheep and deer.

ANCIENT OAKS

Ancient oaks have survived in our landscape not only because they have always been useful for their regular crops of timber but also because they have a place in the culture and the hearts of the British people. Take a glance through virtually any book on trees, from John Evelyn to the present day, and you will find references to famous, old oaks at some point. In their book *Oak – A British History* Harris and James list a staggering 736 named oak trees from across the country. Admittedly, some of these have now disappeared but the legends survive.

The names of some of these trees connect to the places where they grow (or grew) and some refer to those who owned or planted them. Many are dubbed 'great' in honour of their impressive size. Britain's greatest English oak is a mighty specimen with a girth of 42 feet at Bowthorpe Park Farm, near Bourne in Lincolnshire. There is a legend from the eighteenth century about how one of the squires had benches and a table installed inside the hollow trunk so that he could dine there with 20 friends. It's certainly a massive tree but the story must surely be something of an exaggeration.

The other most common associations are with royalty and you can find 'king' and 'queen' trees all over the place. Monarchs may indeed have owned some of them or paid them a special visit but many of them would have been planted by local communities in celebration of a coronation or a jubilee. Elizabeth I's name crops up quite a lot but many of the stories are probably apocryphal. Some of the oaks linked to her are just not old enough to qualify, though it is known that some trees were replaced with seedlings grown on from their own acorns, so it is possible that they are direct descendants.

Queen Victoria travelled widely throughout her realm during her long reign and it is said that she had many favourite oak trees and planted quite a few herself. Because she survived to celebrate not just a golden jubilee but also a diamond one she left her legacy in the hundreds of oaks that people planted in her honour all over the country, and the custom has continued. When Elizabeth II was crowned in 1953 the nation was once again swept up with patriotic fervour and planted 'coronation' oaks in commemoration. A rather fascinating turn of events in the Forest of Dean connects the reigns of all three queens. In 1861, Queen Victoria's husband, Prince Albert, planted what became known as Prince Consort's Oak, near Speech House. This tree had been grown from an acorn taken from the Panshanger Oak in Hertfordshire, which was, allegedly, planted by Elizabeth I. In 1957, Elizabeth II visited the Forest of Dean, where she planted another oak tree, this time one grown from the Prince Consort's Oak, thus completing the link. Happily, all three trees are still thriving. The Panshanger Oak, originally planted on what was once the estate of Earl Cowper, was, as the agriculturist Arthur Young recorded in 1804, 'a most superb oak, which measures upwards of seventeen feet in circumference at five feet from the ground. It was called the Great Oak in 1709.' Various authorities measured it throughout the nineteenth century and the English naturalist and artist Jacob George Strutt made a sketch of it in 1822 for his book *Sylva Britannica* (1826, reprinted in 1830). The tree's shape today compares well with Strutt's drawing and it currently has a girth of 25 feet, which is exceptional for a maiden oak tree.

Probably the most famous royal association with the oak comes from Charles II's well-documented escape from the Parliamentary forces at Boscobel House in Shropshire after the Battle of Worcester in 1651. The King hid here for several days among the boughs of a big, old, pollard oak in Spring Coppice with his loyal follower Major Careless. Apparently they were so uncomfortable in the tree that the two men stole back into the nearby house to sleep at night in a secret attic. Eventually Charles escaped to France. He returned to England in 1660 and on the day of his arrival in London, 29 May, his supporters wore a sprig of oak as a sign of their allegiance to him. This date is now known as Royal Oak Day or Oak Apple Day and the oak has been a royalist symbol ever since, giving rise to all manner of artefacts – documents, heraldry and coinage – and a proliferation of Royal Oak public houses.

The Boscobel Oak became a place of pilgrimage for the royalists, who would carve bits off the tree to keep as sacred relics. This practice probably killed it off altogether eventually and John Evelyn noted in 1706 that what was left of it was dead. By the early nineteenth century the writer Revd C.A. Johns noted in his Forest Trees of Britain that the present Boscobel Oak (assumed to have been grown from an acorn of the original tree) was then in a very dilapidated state, and it is most likely that this also perished. Another oak now grows close by the spot and people like to say that it sprang up from an acorn from the original tree, although it was probably from the second generation tree. This tree suffered a lightning strike in 2001 and so the following year another one was planted by Prince Charles, using a sapling grown from an acorn from the current tree so this time the provenance is certain.

In Scotland a similar event had occurred some 350 years previously when William Wallace and 300 of his men hid from the English in an oak tree near the town of Elderslie. Strutt sketched this tree too, giving the fullest of details of its dimensions. Like the Boscobel Oak, it suffered from the overzealous attentions of souvenir hunters but it was a storm in 1856 that finally killed it.

The Boscobel Oak in Shropshire. Far too small to be the oak tree where Charles II hid from the parliamentary forces in 1651. Storm damage in 2010 caused the loss of the bough on the left. This tree is reputed to have grown from an acorn from the original tree.

Heroes and villains of all description have links to oaks. Beneath a giant oak (now gone) at Keston in Kent , William Wilberforce petitioned the prime minister, William Pitt the Younger, for support with his bill to abolish slavery. The famous Billy Wilkins' Oak at Melbury Park in Dorset commemorates the estate bailiff who ran to warn his master, Sir John Strangeways, of approaching Parliamentary forces during the Civil War; he was overtaken and killed before he could deliver the news. There are oaks beneath which the poets Dryden, Pope and the bard himself, William Shakespeare, are supposed to have penned many of their fine works.

Folk hero-cum-rascally highwayman Dick Turpin had several trees named after him and, reflecting the punitive measures awaiting such criminals, there are many oaks with gloomy reputations as hanging trees. Andrew Morton in his *Trees of Shropshire* tells of the Bolas brothers, notorious highwaymen who were gibbeted on an old oak in the village of Wellington in 1723: 'As was the custom they were nailed to the tree and their bodies were left until they literally fell off the trunk in pieces.' One of the most bizarre stories connected with a hanging tree concerns the Wilverley Oak in Hampshire. Legend has it that just as one poor miscreant was about to be strung up he was struck by lightning. This act of God blew off all his clothes. Whether this killed him before he was hanged or whether it was a celestial sign of his innocence is not recorded but after this incident the tree was known locally as the Naked Man or the Tree of Good and Evil.

One of the most sinister of these hanging trees is the Wyndham's Oak near Silton in Dorset, which was named after a cantankerous and unpopular judge, Sir Hugh Wyndham, who lived close by during the seventeenth century. It is said that he would sit beneath the tree in contemplation. During the backlash to the Monmouth Rebellion of 1685 the infamous Hanging Judge Jeffreys condemned two local supporters of the Monmouth cause to be hanged from this tree, thus consigning it to its gory place in history.

Ancient oaks have had some very strange uses and associations. The massive bole of the Greendale Oak in Nottinghamshire was subjected to monstrous vandalism by the Duke of Portland in 1724. In order to win a wager, he had a massive archway some 6 feet wide and 10 feet high cut through the middle of this tree, which had a girth of 35 feet, so that a carriage and four horses might be driven through. The oak survived its butchery until the late nineteenth century. A bizarre story attends the hollow Arley Oak in Cheshire, where a former squire of Arley Hall once squeezed 11 boys inside the tree and paid them a shilling each for taking part in the experiment.

Oaks have always been singled out as special and in rural communities they have often played a role in religious ceremonies, such as weddings and the beating of the parish bounds at Rogationtide. When people gathered together in the open air to listen to a preacher the village oak was often the designated meeting point. These trees were known as gospel trees or preaching trees. The Meavy Oak in Devon was a gospel tree as well as a focal point for celebrations. In the late nineteenth century The Revd Baring-Gould described how the tree was prepared for the annual village festival. The villagers clipped the top of the tree flat and then erected a platform over it and furnished it with tables and chairs where feasting and entertainment took place.

THE USEFUL OAK

When John Evelyn published his *Sylva, Or A Discourse of Forest-trees* in 1664 he criticized irresponsible landowners for the demolition of 'goodly woods and forests' in favour of a 'disproportionate

The Greendale Oak, as depicted in the 1st Hunter Edition of John Evelyn's Silva.
A tree that was 'modified' in order to win a wager.

The Marton Oak, in Cheshire, is arguably the biggest sessile oak in Britain with an impressive girth of 44 feet. Its massive bole, fragmented into three sections, has looked much the same for over 200 years, as an engraving from 1810 attests. The tree is generally thought to be about 1000 years old, but may be considerably older.

spreading of tillage'. He urged them to protect the oaks in order to support of the nation and its naval defences. He was not the first to voice these concerns and he certainly wasn't the last.

The native oak has always played a central role in the history of British shipbuilding. Detailed records from the Admiralty reveal exactly how a first-rate ship of the line was built, what materials were needed and what it all cost. The building of Nelson's flagship, HMS Victory, which rests today in Portsmouth dry dock, is particularly well documented. Chatham Dockyard received the warrant for her construction in 1759, she was launched with little ceremony in 1765 and finally commissioned in 1778. It took the timber of 6000 trees (roughly equating to 100 acres of woodland) to build her. Some 90 per cent of this was oak and the remainder was made up of elm, probably for the keel, with fir, pine and spruce selected for the masts and yards. She cost a massive £63,176 to build, which is an equivalent cost to building an aircraft carrier today. At 226 feet long and 52 feet wide she's an impressive sight, and in her 32 years of service she was renowned as one of the fastest first-rate ships of the line, with excellent handling. By the end of the eighteenth century international trade was beginning to demand bigger and faster merchant ships and there was plenty of money around to fund the building of new vessels. Wooden ships continued to be built until well into the second half of the nineteenth century, when iron hulls began to eclipse timber construction. Wood carried on being used for fishing vessels, barges and pleasure craft into the nineteenth century but with a few exceptions this material has been cast aside in favour of steel and glass fibre.

Of maritime purpose, yet firmly rooted on shore, was John Smeaton's splendid design for the Eddystone Lighthouse, to the west of Plymouth, which was completed in 1759. Smeaton conceived his design after studying the form of the oak tree 'which, though subject to a very great impulse from the agitation of violent winds, resists them all, partly from its elasticity, and partly from its natural strength'. As a late nineteenth-century commentator noted: 'how wisely he acted in choosing Nature for his instructress.' Smeaton's lighthouse survived until 1870, when cracks in the rocks on which it stood began to cause concern. It was dismantled and rebuilt on Plymouth Hoe, where it stands to this day as a monument to a great designer inspired by the oak.

Oak has always served the needs of builders on land too. From the grandest buildings right down to the humblest cottage or barn, oak has always been the preferred timber. For the majority of timber-framed buildings the oak was traditionally cut and used while still green (unseasoned), and because of the problems of transport most regions of Britain with a good representation of timber vernacular are to be found where there are large concentrations of oakwoods and/or where there is little suitable building stone. However, from the sixteenth century onwards, where suitable clay was available, the increase in brick manufacture changed some of these trends. The many regional variations in design and artistic detail are a fascinating legacy for architectural experts to study. The main thrust of the construction is divided between cruck-framed buildings, which are built around massive pairs of cruck blades at either end and, in larger buildings, also set at intermediate stations through the construction, or box-framed buildings, with their squares and rectangles forming interlocking boxes and creating a superbly rigid framework. Often frames and trusses for buildings were made up off site in workshops or even in the woods themselves, where massive timbers were sawn over great sawpits with huge two-man saws, one man on top of the log and the other in the pit. The frames would be fitted together and all the joints marked by the carpenter, so that when the building was brought

Lord Nelson's mighty flagship HMS Victory *at Portsmouth.*

to site all the pieces would fit together properly once again, held together with oak pegs. Carpenters' marks can still be seen today inside old houses.

Within many larger old houses some beautiful examples of carved-oak panelling, banisters, newel posts and contemporary furniture, sometimes dating back to the sixteenth century, can be found. This glorious old oak has a deep, dark colour, its distinctive patina acquired over centuries of polish and human contact. Carvings in churches and cathedrals are also very special; pew ends, pulpits, rood screens and choir-stalls with their saucy misericords were all invariably of oak. The convergence of oak and the wood-carver reaches its mystical apotheosis in the stirring images of the Green Man, the pagan decorative carved image, as he so frequently glares down from his wreath of oak leaves or suffers them to burst forth from eyes, nose or mouth.

The most valuable commodity ever extracted from oakwoods was tan bark. Huge quantities were required by the tanneries,

Women's work in 1835. A fascinating engraving from The Penny Magazine *shows a 'gang' of women peeling tan bark from a huge oak tree. The women are industrious, while the gentleman foreman surveys.*

particularly from about 1780 until the mid-nineteenth century, for the tannin necessary to make fine leather. Huge tracts of coppiced oakwoods were managed on regular rotations to satisfy the demands of this flourishing industry. In the spring, when the sap was rising, great gangs of men and women would head into the woods, the men to cut and carry, the women to strip the bark with distinctive spoon-bladed knives called barking irons or peeling irons. The coppiced oakwoods also provided a ready supply of wood for making charcoal, which was needed in great quantities for iron smelting and, until the mid-twentieth century, for filtration and insulation purposes. As modern fuels and chemical substitutes made these woodland industries redundant, many vibrant oakwoods slipped into unmanaged neglect.

THE OAK OF GOOD OMEN

The oak was sacred to the Greeks and the Romans, being associated with Zeus and Jupiter respectively, gods of sky, rain, lightning and thunder, and in Norse mythology this was transposed to their god Thor. People worshipped these gods in order to bring fertility to the land and ensure good crops. The pre-Christian Druids worshipped among oak groves and sought out those rarest of oaks, those that bore mistletoe, deemed by them to be a sacred plant bestowed by the gods, as it had no link with the earth. As long as it never touched the ground when harvested it would retain its magical and healing powers. Felling a mistletoe oak has always been considered a most terrible deed, and disaster will befall anyone who does this.

The widespread use of the oak leaves and acorns within old churches harks back to pagan veneration. Acorns were thought to offer protection against lightning strikes and they were often carved on stair banisters as architectural finials or as the tiny toggles for pulling blinds. People used to carry acorns in their pockets for this reason too and to ward off illness and to ensure fertility, potency and longevity. Medicinally the oak offers several beneficial cures. Taken in a variety of forms (ground acorns, bark or decoctions from leaves) oak has long been used to combat irritations and inflammations of the digestive tract. The astringent action derived from oak bark relieves respiratory congestion, cuts and grazes and tones blood vessels and muscles.

OAK INTO THE FUTURE

When the first edition of this book was published back in 2006 it seemed as if Sudden Oak Death (SOD) was likely to be the main concern for oak tree health, but over the ensuing years that perception has changed.

Even though SOD had wrought great losses on the oaks in America it transpired that *Phytophthora ramorum*, the deadly pathogen responsible, did not have a taste for the two indigenous British oaks. Instead it turned its unwanted attentions on American oak species growing in the UK, as well as beech, rhododendrons, viburnums and latterly the Japanese larch. The devastation of thousands of hectares of commercial larch plantations has cost British forestry dearly.

Instead, over the last ten years or so the main threat to British oaks is now thought to be Acute Oak Decline (AOD). The disease is believed to have been in the UK oaks since the late 1980s, but has only escalated its spread and intensity in recent times. Caused by a group of harmful bacteria (*Brenneria goodwinii*, *Gibbsiella quercinecans* and *Rahnella victoriana*) the diseased oaks display black bleeding lesions from bark fissures, beneath which the vascular tissue decays and inhibits transmission of water and nutrients to the crown. The

Hoar-frosted oak in the Lodon valley, Herefordshire.

foliage dies back and trees usually die within four to six years. The vector that carries the bacteria through the tree is thought to be the larvae of the two-spotted oak buprestid beetle (*Agrilus biguttatus*), although other factors may also be responsible. There is no cure!

However, it's not a completely bleak outlook. Some trees seem to exhibit resistance to the AOD and even many of those that do succumb can later recover and regenerate. Generally, the overall future prospects for oak in Britain are good, as it is a native species at the northern limit of its range and climatic changes should suit it well.

The monarch oak, the patriarch of trees, Shoots, rising up,
and spreads by slow degrees; Three centuries he grows,
and three he stays Supreme in state, and in three more decays.

FROM 'THE MONARCH OAK' BY JOHN DRYDEN (1631–1700)

The most famous oak in Britain is indisputably the Major Oak in Sherwood Forest. For at least 200 years visitors have been lured here largely through the romantic tales of Robin Hood and his cohorts. Here is a Sunday school group dressed in their best for the outing around 1895. Today some 350,000 visitors visit the tree every year.

Ash

The common ash (*Fraxinus excelsior*) has a huge range. It grows all over Europe, from the Caucasus in east and as far north as Norway. In Britain it spreads up into the northwest corner of Scotland. It is one of our most adaptable broadleaves, able to colonize more successfully than most and to thrive in some incredibly inhospitable situations.

The ash tree towers; ancient and friendly arms
Stretch out on all sides into the gracious sky;
Its leaves like hands gesture continually;
It woos the blue airs with compulsive charms.

FROM 'THE ASH TREE' BY WILFRED CHILDE (1890–1952)

A delicate spray of ash flowers.

One of the ash's most important contributions to the landscape has been its colonizing of hedgerows in areas blighted by storms, such as those of 1987 and 1990, or denuded of elms in the wake of Dutch elm disease. This makes it an extremely valuable tree for landscape character, shelter and wildlife. Ironically the tree that helped to heal the voids created by Dutch elm disease is now itself threatened by the inexorable spread of ash dieback.

The ash is fast-growing and highly resilient when coppiced or pollarded and the timber is strong and bends well when steamed. Traditionally used for agricultural implements, it was often known as the husbandman's tree. It is also ideal fuel wood and burns well even when green. The ash is one of the most graceful of Britain's native broadleaf trees and its delicate beauty has given rise to an equally beautiful name: Venus of the Woods. It is light and airy, with fine sprays of pinnate compound leaves, which have six to twelve pairs of unstalked, elliptic-ovate shallow-toothed leaflets. The tree produces its flowers before its leaves and has one the shortest seasons of all the broadleaves, being one of the last to show its foliage (sometimes not until early June) and the first to fall. For this reason, and also because of its relatively dull autumnal colour, some consider it to be a poor choice for the garden. The Scottish writer Sir Thomas Dick Lauder certainly thought so when he wrote in 1834: 'Ash trees should be sparingly planted around a gentleman's residence, to avoid the risk of their giving to it a cold, late appearance, at a season when all nature should smile.' The silvery-grey bark of the ash remains quite smooth for the first 30 or 40 years of its life, only gradually developing its characteristic fine, evenly fissured texture. The tree is clearly defined in winter by its tight sooty-black buds and by its drooping or arching boughs, which are upturned at the end.

Alan Mitchell describes ash flowering most succinctly as 'total sexual confusion'. Trees may produce male, female or hermaphrodite flowers, either on separate trees or all together on individual trees (trioecious or subdioecious). Studies carried out in 1991 revealed that there tend to be more predominantly male trees than females, and when hermaphrodite flowers are part of the array they manifest in three types: either purely hermaphrodite or alternatively with vestigial male or female organs, these latter two types still having the ability to function as either males or females or even as both male and female. Equally, some trees that bear predominantly male or female flowers will not necessarily produce pollen or seed.

Ash trees also have the remarkable capacity to change their sexuality from one year to the next. The reason is something of a biological mystery, but at some point in the evolution of the ash it became necessary for the tree to adopt this strategy. Possible explanations could be climatic variation, increasing the fecundity of the tree when conditions for seed germination seem more favourable, or perhaps some sense of competition (or lack of it) that stimulates the tree to produce more seed when the immediate habitat is threatened by proliferation of rival species or, equally, when space opens up and there are increased chances of colonisation. Horse chestnuts can do this too but no one really knows how or why this occurs.

Common ash usually begins to bear flowers at thirty to forty years of age. The dense axillary panicles of tiny, dark maroon

An impressive ash tree on a Devon hill top.

An ash, colonized by a rowan tree, which has grown from a berry deposited by a bird in the fork of the ash. A sprig from such aerial rowans is believed to be a powerful amulet against the forces of black magic.

or purple flowers, devoid of petals, on the shoots of the previous year's growth appear before the leaves flush, usually by mid April, but it can be as early as late March or, conversely, well into May. This results in communities of trees, some of which have barely broken bud, alongside others that are in full leaf and already forming clusters of fruits. The assumption is that this is a survival mechanism; the trees genetically programmed to flower over an extended period to allow for damage to the earliest inflorescence from late spring frosts. The male flowers appear first, resembling little clumps of dark coral and when fully opened bear clusters of twin-branched stamens. Hermaphrodite flowers appear next, and lastly the female flowers, each having a style tipped with a dark red stigma. These burst forth in exotic-looking feathery sprays on long green stems, each one a tiny sylvan, pyrotechnical explosion. The females eventually develop into pendulous bunches of bright-green winged fruits, known as samaras or more commonly keys, which contain the seeds. Anyone who has spent time trying to weed seedlings out of their garden will appreciate how well they germinate. I was once rash enough to offer one of my daughters a penny for every seedling she pulled from the flower beds. She cleaned me out.

ASH IN THE LANDSCAPE

When left as an open-grown tree in a natural setting the ash will rival any other broadleaf specimen for grace and beauty. Jacob George Strutt captured its charms perfectly in his beautifully illustrated book *Sylva Britannica*:

> It is in mountain scenery that the ash appears to peculiar advantage; waving its slender branches over some precipice

> which just affords it soil sufficient for its footing, or springing between crevices of rock, a happy emblem of the hardy spirit which will not be subdued by fortune's scantiness. It is likewise a lovely object by the side of some crystal stream, in which it views its elegant pendent foliage, bending, Narcissus-like, over its own charms.

The species does well in many different settings. When the trees are planted closely together in a woodland they will grow up straight very fast, forming long, whippy stems; most of the side shoots will die back for lack of space and light. As a maiden tree the ash will seldom exceed 150 years, its timber reaching maturity and ready to be harvested in 50–70 years, after which time it often begins to rot from the heart. Careful management of these trees over centuries for their timber has been a life-insurance policy for many old ashes.

Ash is not noted for its longevity, but this huge ancient pollard in an old deer park may well be over 500 years old. With a girth of over 29 feet it's a contender for Britain's largest ash.

An ash pollard in the Black Mountains, on the Welsh border.

Ashes are often found as coppice stools or pollards in woodlands or hedgerows and there are examples in Britain today that are reckoned to be more than 1000 years old. In many places these ancient stools have been laid and re-laid repeatedly and allowed to run for great distances. Hedge layers will sometimes leave a few upright stems to grow vertically and these tend to develop into small, linear woodlands from a single tree. Pollards are useful in hedgerows or pasture land where any new growth might be vulnerable to browsing livestock. Animals will still be tempted to have a nibble at ash bark but these trees are pretty hardy and they generally survive by growing scar tissue to protect themselves. We are used to hearing about ancient oaks and the part they play in British folklore, but the single, open-growing ancient or veteran ash has never quite achieved the same iconic status. One exception is the Clapton Court Ash in Somerset. Clearly ancient, but age unknown, it measures a monstrous 29 feet around its girth at chest height. The mass of gnarled, scar tissue at the base of the trunk, peppered with epicormic shoots, resembles a moss-covered boulder – an indication that it has probably suffered from frequent depredations over a long period by browsing animals. This tree is growing in a private garden and is not accessible, but luckily this is not always the case. Talley Abbey, near Llandeilo in Carmarthenshire, boasts a huge ash that has been growing in excess of 200 years in an old hedgerow a short walk from the main building. Numerous burrs, boughs and buttresses make this tree a statistician's nightmare. At a height of about 6 feet it measures 25 feet around its girth; halfway between this point and the ground it is an impressive 36 feet.

An unusual variety of the ash is the weeping form 'Pendula'. This graceful tree was first noticed as a sport or natural mutation of common ash in around 1760, where it was found growing in a field belonging to the Vicar of Gamlingay, near Wimpole in Cambridgeshire. Much scion-wood was taken from this tree and grafted on to common ash stocks, as the seeds would not have reproduced this form. Records show that this same sport has occurred naturally from time to time but most trees will have been propagated.

Ash is a demanding tree, drawing in water and nutrients from a wide area through a remarkably extensive and fibrous root system. This has led to the traditional belief that it is unlucky to plant anything else within its range and there is some truth in this: young seedlings will not thrive in its shadow. Wild flowers are a different case, however, and those that come out in the early spring, such as snowdrops, wood anemones and bluebells, can take advantage of the light before the ash comes into leaf.

The ash is a common feature in many mixed-species broadleaf woodlands, especially on chalk or limestone. In southern England

The unusual weeping form of ash.

the downland that also favours the yew often sees the ash in a complementary role. It is an ever present element across the Mendips in Somerset, although botanist and conservationist Peter Marren believes that its dominance over the hazel and the small-leaved lime may be a relatively recent development. Travel to the north of England, to the limestone of the Derbyshire and Yorkshire dales and there is no confusion – this is true ash country indeed.

Noted lime expert Donald Pigott has established that the presence of huge, old lime coppice stools in Derbyshire woods indicates that in warmer times, perhaps thousands of years ago, the lime was an important element. A cooler climate, which has seen limes lose their ability to set viable seed, coupled with predations by ever increasing introductions of sheep (which are quite partial to lime leaves), has caused this species to decline. This has left a void that the ash has been only too eager to fill. Its ability to produce large quantities of seed, many of which will germinate in rocky places, inaccessible even to the most agile of sheep, has probably been its salvation.

In North Yorkshire the ash faces its sternest test on the limestone pavements above Ribblesdale, upper Airedale and Wharfedale. These are inhospitable places. The limestone grykes (hollows carved out by rainwater over millions of years) provide the only shelter for a tree from marauding sheep, gales and storms, ice and snow. Alongside a few yews and hawthorns, the ash trees seem to be able to handle these harsh conditions, though they pay a price. Most of them are stunted and they look like bushy little bonsai trees, with their heads tucked well below the parapet.

Further up the valley, and slightly less exposed, are the nature reserves of Colt Park, on an isolated outcrop, and Ling Gill, in a steep ravine. Here the ash is dominant, with many splendid old trees, whose relatively modest size disguises their true vintage, for in such locations growth rates are considerably slower than in the lush lowlands. Impressive carpets of wild flowers and many beautiful lichens and mosses bedeck these woods; they are oases in a spartan landscape that, before farmers arrived with their sheep in Neolithic times, might have been typical of much of the surrounding countryside.

ASH DIEBACK

Since the first edition of this book the fortunes of ash have changed dramatically. Although it had probably been in the UK since the mid 1990s, ash dieback was only officially notified in 2012 and a ban on trade and movement of all ash species put in place. Ash dieback is caused by the fungal pathogen *Hymenoscyphus fraxineus*, originally described and named *Chalara fraxinea* in 2006, hence the commonly used appellation *Chalara.* Trees become infected when windborne fungal spores land on the leaves, damaging them with the toxic chemical viridiol. This then enters the tree, constricting and killing the vascular tissue so that the tree can no longer draw nutrients and water to the crown. Ultimately the tree will die over several years. Introduction of the disease has been put down to two main vectors that are responsible for the initial spread of ash dieback – windborne spores blowing in from Northern Europe and the importation of diseased saplings from nurseries in Europe. Symptoms begin with areas of foliage turning brown, then black; shoots start to die with brown or purplish brown necroses, sometimes with characteristic diamond lesions around the base of sideshoots. Branches die back, foliage thins, usually displaying a 'pom-pom' effect where the tree is fighting the disease by producing defensive clusters of leaves from lateral buds. As the tree dies it becomes more susceptible to attack from aggressive fungi which will weaken the roots system and root collar making the tree unstable and dangerous. There is no cure for ash dieback and projections suggest that up to 99% of the nation's 150 million mature ash trees will die over the next 15-20 years. And it's not just ash dieback. The future for the tree may also be threatened by emerald ash borer (*Agrilus planipennis*), a beetle that has already killed millions of North American ashes, and is currently making its way around the globe. It hasn't been seen in Britain yet, but its arrival is almost inevitable.

One ray of hope is that a handful of trees have already shown a degree of tolerance to the disease. The Ash Archive, the culmination of five years of research initiated by The Living Ash Project, is a collection of 3,000 ash trees recently planted in Hampshire comprising cuttings taken from ash dieback tolerant trees observed in the wild and grafted on to ash rootstocks. The aim is to monitor the development of these trees, learning whether or not they have an enduring tolerance to the fungus, with the ultimate goal of producing seed orchards derived from these trees that will make it possible to plant tolerant trees back in the wild at some point in the future. With the epidemic spreading and the certainty that at some point in the future mature ash trees will be a scarce sight I was moved to produce a monograph in 2018. *ASH* explains ash dieback in greater detail, but is also a complete portrait of the tree's contribution to the history, culture, economy, folklore, ecology and landscape of Britain.

THE USEFUL ASH

Timber from the ash is very versatile and John Evelyn comes to the fore once again with an admirable account of its virtues:

> The use of Ash is (next to that of the Oak itself) one of the most universal: it serves the soldier and heretofore the scholar, who made use of the inner bark to write on, before the invention of paper. The carpenter, wheel-wright, and cart- wright find it excellent for plows, axle-trees, wheel-rings, harrows; it makes good oars, blocks for pullies, and sheffs, as seamen name them. For drying herrings no wood is like it, and the bark is good for the tanning of nets; and like the Elm, for the same property (of not being apt to split and scale), is excellent for tenons and mortises; also for the cooper, turner, and thatcher; nothing is like it for our garden palisade-hedges, hop-yards, poles and spars, handles and stocks for tools, spade-trees, &c. In sum, the husband-man cannot be without the Ash for his carts, ladders, and other tackling, from the pike to the plow, spear, and bow; for of Ash were they formerly made, and therefore reckoned

Ash the survivor. This tree seemingly grows from the bare limestone pavement above Ribblesdale in Yorkshire.

> amongst those woods which, after long tension, has a natural spring, and recovers its position, so as in peace and war it is a wood in highest request.

The landscape designer J. C. Loudon recommended ash wood for kitchen tables because it tends not to splinter when it is scoured and the naturalist and artist Prideaux John Selby noted that five- or six-year-old coppice wood was very suitable for ashplants (walking sticks) and barrel staves. From the craftsmen of old, who built carriages and carts, to the present-day manufacturers of specialist motor cars like the Morgan, ash has been the timber of choice. (See also Introduced Broadleaf Trees.) Because it is remarkably flexible it can be bent and formed (at one time through immersion in water but latterly by steaming) while maintaining its integrity. With the advent of manned flight, ash was often favoured for the highly stressed elements in the construction of aircraft frames.

Sadly, most of the old, traditional uses for ash wood are now obsolete, mainly because plastic and metal have taken over, but there are still a few obscure, niche markets left. Walking-stick makers still prize the young coppice produce and green wood-workers are fond of ash too, as it works well for furniture- making, whether it's turned on the pole lathe, pared down on the shaving horse or steamed for hoop-backed or balloon-backed chairs. Morris dancers choose ash sticks to thwack with gusto, as they're tough and unlikely to splinter. And for the same reason, there is a ready market for quality ash butts from which to make hurleys – the large spoon- ended sticks that the players use to wallop the ball in their traditional game of hurling.

Even without all the wonderful things that can be made from ash, it has always been regarded as the best fuel for the fire as it burns bright, with a good flame, little smoke and plenty of heat.

Constructing the ash frame of a Morgan sports car in the Malvern factory. Each craftsman can build three to four of these frames in a week.

Evelyn called it 'the sweetest of our forest fuelling, and the fittest for ladies' chambers' and the final lines of an anonymous poem called 'Logs to Burn' endorse the sentiment:

But ash logs, all smooth and grey,
Burn them green or old;
Buy up all that comes your way;
They're worth their weight in gold.

MYTHS, LEGENDS AND SUPERSTITIONS

With such a universal utility and geographical range, it's hardly surprising that ash figures prominently in British place names, especially in England and Wales. Cambridge botanist Oliver Rackham tells us that from Anglo-Saxon times onwards only the thorn features more frequently. There has been a long-standing tradition of mystical, mythical and medical associations with ash, much of which must have blown in with the Nordic invaders. Most of these are of good omen so it seems logical that settlements with the ash name may well have deemed it spiritually auspicious in some way.

Ash is the World Tree, Yggdrasil of Norse mythology, from where the gods created the universe and the first man and woman – 'Ask' from the ash and 'Embla' from the elm – and from which Odin hung himself in his search for eternal knowledge.

Even ash, even ash
I pluck thee off the tree;
The first young man that I do meet
My lover he shall be.

This anonymous verse is just one of many old country sayings that claimed to tell a young woman who her husband or sweetheart would be. She was supposed to put the leaf in her left shoe and wait for the prediction to come true.

Ash has long been held to be of great efficacy against many ills and evils. Pliny the Younger wrote of how ash leaves 'are of so great a vertue against serpents, as that the serpents dare not be so bolde as to touch the morning and the evening shadowes of the tree, but shunneth them a farre off'. He also wrote that serpents would prefer to creep through flames rather than cross a circle of ash leaves. Nicholas Culpeper was moved to refute this latter assertion as superstitious tosh and it seems he had put the theory to the test, finding 'the contrary to which is the truth, as both my eyes are witnesses'. However, he still held with the common claim, first espoused by the first-century Greek physician and botanist Dioscorides, that the juice of ash leaves 'taken inwardly, and some of them outwardly applied, are singularly good against the bitings of viper, adder, or other venomous beast'. Culpeper also considered that an extract from the leaves would 'abate the greatness of those that are too gross or fat'.

Evelyn reported that 'There is extracted an oil from the ash which is excellent to recover the hearing, and that for the tooth-ache the anointing of the offending teeth therewith is most sovereign.' As a culinary aside he also recommended pickled green ash keys as a salad delicacy.

An interesting account comes down from The Revd John James Lightfoot, botanist and chaplain to the Duchess of Portland, in his *Flora Scotica* of 1777. In his day it was common practice in parts of the Highlands for the nurse or midwife to put one end of a green ash stick into the fire when a baby was born and give the newborn a spoonful of sap from the stick, the assumption being that the child would absorb the strength of the ash and be protected from witches and goblins. This belief in the restorative powers of the ash was common in rural areas until the middle of the nineteenth century. The English eighteenth- century diarist and naturalist The Revd Gilbert White of Selborne, writing in 1776, describes a row of pollard ash trees that he knew of that had been split with an axe. The custom was for parents to pass a child naked through the trunk of an ash tree in order to heal a rupture. This would only happen when the tree was bound up again and recovered. This strange ritual has also been recorded in Devon, one instance occurring as late as 1902.

It used to be a popular superstition among country folk that shrews could harm their livestock and they thought that if an animal were sick or lame then the damage must have been caused by a shrew running over it in the field. So many communities would have a 'shrew-ash': they would place a live shrew inside a hole that they had bored in the tree's trunk and then seal in the poor creature to await its fate. This ritual had the effect of consecrating the tree and so it acquired magic powers. People would then cut a branch off and stroke their stricken beasts with it to effect a cure. There are many accounts of this custom, some of them going into great detail about the mechanics of the process. Success rates, however, are poorly recorded.

Modern medicine is more pragmatic but science has shown that some of the old claims made for the healing properties of the ash have some foundation. Ash is known to have diuretic and laxative properties that can benefit the gastric system. It has been used to treat gout, rheumatism and dysentery and to stem bleeding and has served as a quinine substitute in the treatment of fevers and malaria.

Oak before ash, we're in for a splash;
Ash before oak, we're in for a soak.

This charming old rhyme, so frequently quoted, suggests that oak and ash leafing can predict how much rainfall the summer will bring but it appears to be an old wives' tale. In 2005 environmental scientist Dr Tim Sparks decided to check this out. Luckily for him, the Marsham family in Norfolk had been keeping a continuous record for 158 years. He found that in all that time ash had beaten oak on only 46 occasions. He checked the rainfall records for each year from May to September and discovered that no matter which tree leafed first, the average monthly rainfall across the board was virtually the same. However, over the last 42 years ash has beaten oak on only three occasions and Dr Sparks puts much of that down to the general warming of the climate, which has caused the oak to advance its leafing at a greater rate. The warmer we get, it would seem, the less likely it is that this old rhyme will hold water.

Beech

Of all the native trees in Britain the statuesque beech (*Fagus sylvatica*) seems to be the most accomplished at exhibiting resolute strength counterbalanced with a grace and delicacy of form. This then is the Mother of the Forests, the Madonna of the Wood or, as the Scottish writer George MacDonald in his *Phantastes* calls it, 'the White Lady, who saved a man from the malice of an ash'. The tree is particularly striking in winter, when one can appreciate its smooth silvery-grey bark and the fine tracery of its boughs and twigs.

Since youthful lovers in my shade
Their vows of truth and rapture made,
And on my trunk's surviving frame
Carved many a long-forgotten name.
Oh! by the sighs of gentle sound,
First breathed upon this sacred ground;
By all that Love has whisper'd here,
Or Beauty heard with ravish'd ear;
As Love's own altar honour me:
Spare, woodman, spare the beechen tree!

FROM 'THE BEECH TREE'S PETITION'
BY THOMAS CAMPBELL (1777–1844)

Beech nuts or mast.

Beech is certainly an indigenous species throughout the greater part of Europe, but whether or not it is a true native across the whole of Britain has long been debated. Reliable pollen records only go back about 3000 years, with its presence at that time confined to the southeast corner of England. Some authorities believe that the species may well have been around for about 8000 years, others think 5000 years, and some claim that human introduction occurred some time during the pre-Roman era. The general consensus now is that the beech was a later colonizer than many other species after the ice melted and it is probably safe to say that its natural range in Britain is south of a line between the Wash and the Severn estuary, with a small outlying province in southeast Wales.

The name 'beech' derives from the Anglo-Saxon *boc*, *bece* or *beoce*, the early German *buche* or the Swedish *bok*. This is from where we get our word for 'book' because early books were either made from inscribed tablets of beechwood or else the vellum leaves (again, a tree reference – 'folio' from the Latin *folium*, a 'leaf') were bound between beechen boards. The inclusion of 'buck' in many English place names indicates the presence of beech and the best-known example of this is Buckinghamshire. This county is still famous for the beechwoods that spread across the Chilterns, though these have been selectively managed over the last two centuries and can no longer be considered to be natural woodland.

A handsome autumnal beech near Bolton Abbey in Yorkshire.

THE BEECH IN THE LANDSCAPE

Beech will grow in most free-draining locations and it does very well on the chalk and limestone of southern England, principally because the feeding roots draw nutrition from the surface, acid layer. It also thrives on quite acidic or sandy soils, such as those in the New Forest. It has a particularly shallow rooting system, which, for the purposes of support and nutrition, covers huge areas beneath mature trees. This, combined with a dense summer canopy, makes life for any other plants beneath its spread well nigh impossible; only early flowering species, such as primroses or wood anemones, stand a chance before the summer gloom descends upon the woodland floor.

When the fresh lime-green leaves of spring emerge they are so soft that they feel like a delicate fabric and have a fine fringe of

Springtime in a beech plantation. Young trees grow tall and straight – ideal for foresters.

silky hairs. To walk among springtime beeches is to bathe beneath an eerie and diffuse green floodlight. In the landscape the contrasting visual effect of these green beacons when set amongst dark holly or yew is most striking. The beech comes into its own again in autumn with an infinite array of yellows and oranges, creating a glorious glow. It tends to retain many of its leaves throughout the winter, when young or trimmed into hedges.

The commercial need for beechwood in the past has meant that many woodlands were managed to produce fuel wood or charcoal. In the Chilterns the demands of the furniture industry based around High Wycombe during the nineteenth and early twentieth century kept them flourishing. Regular coppicing or pollarding helped to extend the life of the trees and maintain the woodlands' vibrancy and habitat diversity. The risks for beeches today in woodlands is that once they are no longer commercially useful they are either clear-felled and replaced with something that is faster growing, such as conifers, or they are allowed to grow on regardless and die out naturally.

Various beechwoods across the country have their own specific qualities and reflect many different aspects of management – or lack of it. These woods may be missing the more attractive gnarled, old trees but the ranks of stately straight stems have a certain uniform majesty as they thrust upwards for the light. For most foresters this is what beechwoods are all about – close-grown, straight-timber trees.

BEECH MANAGEMENT

In the past beeches were generally coppiced but this is not always successful in modern forestry, as for many beeches this signals their demise. The naturalist Richard Mabey feels that we have forgot-

An ancient beech pollard on the Blorenge, near Abergavenny.

ten the finer points of the technique; the old foresters probably left one or two major stems to aid the tree in its regeneration process, perhaps coppicing these at a later stage once the new growth was established. This makes a good deal of sense. However, if we have forgotten this, then it is evident that we have also forgotten how to pollard old beech trees. Ash, willow and even the scarce black poplar are all very responsive to pollarding and will regenerate freely immediately after cutting. Not so the beech or the oak, which have largely fallen out of this regime, the knowledge of the process having faded with history.

Helen Read and her team at Burnham Beeches, an old wood-pasture and heath wedged between Slough and the M40, have re-established a pollarding programme for many of their 500 ancient beech pollards as well as initiating the lopping of new pollards from young trees. These are the veterans of the future. At first there was a measure of trial and error, for not all the ancient trees, some thought to be as many as 400 years old, responded favourably to being cut back after more than a century of neglect. Trauma sometimes ensued and occasionally trees died. It soon became apparent that the procedure was most successful using a gradual, staged approach, whereby some old growth was left to sustain the tree until new growth became established. In a perfect world the trees would have been left to their own devices but large boughs on outgrown pollards can become unstable and cause an imbalance to what is essentially quite a small tree beneath. Windthrow of large pollard boughs can also cause substantial damage to the whole tree. Pollarding will usually extend the lifespan of ancient trees, for by reducing their physical structure they are 'fooled' into believing that they have yet to mature; so a beech that might do well to last 200 years as a maiden tree is able to survive for twice as long. Even when it is really ancient the root system and compact canopy will easily support the small frame of a pollarded beech. As the trees age, the heartwood invariably rots away but, as any engineer will confirm, a hollow tube is a much stronger structure than a solid pole.

In 1987 and 1990 beeches bore the full brunt of the ferocious gales that swept across southern England. They succumbed to the assault in great numbers and it was only when their massive yet shallow root plates were exposed that people realized how vulnerable they were. On a positive note, the vast majority of hollow veteran beeches were unharmed, sometimes losing a few boughs at worst. Understandably, after the storms in 1987 everyone was deeply saddened by the tree losses. With many woods decimated and huge landmark trees snuffed out overnight, whole vistas were altered out of all recognition. Predictably, the knee-jerk reaction was to tidy up the mess and replant. By the time the 1990 storm happened, lessons had been learnt and beech was a particularly good tutor. Beech trees planted after 1987 had often not performed very well. Experts took a look at some of the woods of Kent, Sussex and Surrey, where relatively little had been done, and they found that natural regeneration of beech was remarkably well advanced, proving the old adage that nature knows best.

The beech's ability to grow in some truly precarious situations is never better demonstrated than in some of the hollow ways of Hampshire or the ancient green lanes and tracks above the Usk valley in south Wales. Here the old trees seem to balance on a knife-edge, deeply undercut by the erosion of the banks beneath and often propped up on huge, knobbly old roots. This has probably prevented the existing bank sides from caving in altogether, and the gloomy hollows beneath the trees make perfect roosting sites for the rare Bechstein's and barbastelle bats.

In Selborne in Hampshire, home to The Revd Gilbert White, there is a meeting of two very different types of beechwood. Selborne Common was once wood-pasture, used in the past by commoners to graze their livestock; it has been long neglected and yet many of the old pollard beeches are still in evidence. Selborne Hanger is the steep valley slope along the northeast edge of the common, which hangs above the village of Selborne. The trees here are huge, towering specimens and because they were always less accessible to the foresters on such precipitous ground it's thought likely that many have survived from White's day. White wrote in praise of the fine trees above his home as 'the most lovely of all forest trees, whether we consider its smooth rind or bark, its glossy foliage, or graceful pendulous boughs'.

Something of an oddity amongst beeches grows above Selborne Hanger: a large tree with two huge stems that are fused together by a short cross-bough. Whether this is the result of some form of natural grafting is unknown. The temptation is to believe that some forester formed this union when the trees were young, perhaps as a token of love for his young lady. I first came across this tree ten years ago and I have since discovered several other beeches that have been 'reconstructed' in a similar fashion. The most spectacular of these is to be found near Lisburn in Northern Ireland. During a visit to the area in 1787, John Wesley twined two young beech trees together, thus symbolizing the unity of Methodism and Anglicanism. These two trees growing as one are still in extremely good heart today. In 1906 botanist and plant collector H. J. Elwes wrote of two trees quite similar to the Selborne example that he had come across on the Ashridge Estate in Hertfordshire. He called them 'inarched beeches' and was equally puzzled by the phenomenon: 'it [the inarch] passes into the other tree without any signs to indicate how the inarching took place and might almost have been a root carried up by the younger tree from the ground, as it has no buds or twigs on it.'

An Edwardian postcard shows gardeners busy trimming the formidable beech hedges of Bramham Hall in Yorkshire.

Beech seems particularly resilient to cutting and forming. Massive old coppice stools discovered high above the Wye valley must once have been harvested on a regular rotation; more than a century after their last cut they are now great, grey ripples and folds looking more like the rocks upon which they cling than living trees. Around many fields in Devon and Cornwall beech has been cut and laid along the tops of the old hedge banks. The tree's habit of hanging on to its dead leaves throughout the winter make it a good hedging choice, helping to improve protection from the elements for livestock. Beech hedges in domestic settings offer continued protection and privacy through the winter months. Long beech rows as windbreaks are a feature across the top of the Cotswolds, as are the beech clumps that mark the presence of old burial mounds.

Great old beech trees are few and far between, particularly the tallest specimens, which, as many authorities have reported over the last 200 years, were no sooner classified as champions than they were blown down or felled by the forester's axe. Elwes mentions the 'King and Queen Beeches' at Knole Park in Kent that he saw in 1905. The 'King' tree, then thought to be one of the greatest beeches in Britain at 100 feet high and with a thumping great girth of 30 feet, survived until about 1950 (then measured at 32 feet) when it was claimed by gales. No beeches in excess of 30 feet are known in Britain today. There were also 'King and Queen Beeches' at Ashridge Park in Hertfordshire and when Elwes visited these in 1903 he discovered that the 'King' tree had been blown down in 1891; the astounding 'Queen' tree was almost 135 feet high. All the greatest beech trees in Britain at present are massive old pollards, such as those at Burnham Beeches. Huge, outgrown pollards may be found in Epping Forest, at Felbrigg in Norfolk and still within Ashridge Park. Epping contains a most strange form of beech, known to local tree buffs as coppards. These are old coppice stools whose regenerating stems were left to grow quite large before each one was cut as a pollard. This interesting feature is almost certainly a reflection of a change in the grazing regime for livestock in the forest, possibly one that occurred during the mid-nineteenth century, when land grabs by wealthy landowners were squeezing the commoners out of their forestal rights to graze and collect wood for fuel. Once the coppice stools were no longer fenced off they became fair game for sheep or deer so, in the absence of fences, pollards allowed the trees to regenerate unmolested.

The beech is an ideal choice for formal planting schemes. Beech clumps on hill tops are a favourite and they can look spectacular. One of the most famous of these is the Chanctonbury Ring, which Charles Goring planted on his land on the South Downs in 1760. It was a controversial scheme in its day, as many of the local people felt that he was ruining a natural landscape and desecrating the site of an Iron Age fort. Goring's beeches took a terrible battering in the 1987 gale but the site is recovering and some new trees have been planted. Drivers heading for the West Country will spot a marvellous beech clump next to the A30 near Okehampton. Two of the most famous groups of hill-top beeches are the Wittenham Clumps above the Thames valley in Oxfordshire. Planted in the eighteenth century, on what are more properly known as the Sinodun Hills, the Clumps were painted many times by the artist Paul Nash. Beech avenues are a striking feature of many country estates and one of the most splendid (and arguably one of the longest) is the Grand Avenue, which stretches for almost 4 miles through the middle of Savernake Forest in Wiltshire. The 3rd Earl of Ailesbury first planted it in 1723 and since then many trees have come and gone and come again but the visual effect is still impressive.

THE BEECH IN SCOTLAND

If beech is only truly native in southern England then its progress as far as the north of Scotland is largely ascribed to human intervention. Once there, however, the species obviously naturalized readily enough and then began to colonize naturally. As a result, Scotland has some splendid beeches, including one of Britain's finest examples of a layering tree. The ancient outgrown pollard at Kilravock Castle, east of Inverness, is thought to be over 300 years old. Many of its lower boughs have been allowed to arc over and touch the ground and at these points they have put down roots, which have thrown up fresh growth to form a circle of potential new trees around their parent. It is quite normal for trees to try to replicate themselves in this way – limes, sweet chestnuts and horse chestnuts as well as conifers such as the western red cedar or Wellingtonia all have this capacity – but the phenomenon is not all that common because untidy lower branches are often trimmed away before they get the chance to regrow.

Scotland's most spectacular giant beech hedge can be found at Meikleour in Perthshire. Tradition relates that this was first laid out in 1745 and shortly afterwards the estate workers who had done the planting were called away to fight at Culloden for Bonnie Prince Charlie. They never returned and the hedge was left as a memorial to their bravery. This is a record-breaker and with an average height of 100 feet along its 9/91⁄3-mile length it is officially the tallest hedge in the world. It is trimmed and measured every ten years and it takes four people using cherry-picker platforms about six weeks to complete the job. If you're not expecting this giant as you drive along the A93 it will surely take your breath away as it rolls into view. (See also Hedges and Hedgerow Trees.)

THE USEFUL BEECH

The Romans chose beechwood for their ironworks and glass-making. Beech also produces high-grade charcoal, which was traditionally used to make gunpowder. As a fuel it has long been considered one of the best woods, as it generates great heat and burns with a clear flame. Until the latter part of the nineteenth century many of the hearths of London's homes burned bright with beech from the woods bordering the metropolis.

Beechwood may be hard but it does not have the durability of oak or elm and it is more susceptible than both of these to attack by insects such as the furniture beetle. However, the timber has been used with much success in the past for major construction in places where it needed to be immersed in water, such as pilings for bridges and for lock gates and sluices in canals. Several authorities tell of the piles upon which Winchester cathedral was built, in the midst of a peaty marsh, shortly after the Norman Invasion. When these were examined at the turn of the nineteenth century, some 700 years after they had been sunk into the ground, they were found to be absolutely sound.

Beech has always been the wood of choice for furniture makers of Buckinghamshire, most famously for the legs and spindles of Windsor chairs (the seats usually being of elm). The wood cleaves easily and bends well when steamed. A vast industry once thrived throughout the Chilterns, giving gainful employment to the bodgers who set up their pole lathes in many a woodland clearing and turned thousands of chair legs and spindles from the green beechwood. Most of the bodgers were men while women did most of the caning and rushwork. In 1905 journalist Hugh B. Philpott wrote about the cottage industry of chair-making in and around High Wycombe: 'It has been estimated that the output of chairs

Constructing a beechwood chair (probably with a seat of elm) in the early twentieth century.

from Wycombe amounts to seven or eight for every minute of the day and night; in other words, Wycombe supplies in the course of a year the equivalent of a chair apiece for every man, woman, and child in London.' He also reported that '5000 chairs were made for the Alexandra Palace in six days'. These statistics indicate that there was plenty of work to be had at this time but Elwes was also aware that mechanization and timber imports from Canada were beginning to have an impact. When they were not making chairs the workers could produce all sorts of wooden artefacts. The turning of beech was a complementary skill and there was always a ready market for bowls and other kitchen utensils, such as spoons and rolling pins. Spalted wood (wood that has random black veins running through it caused by disease) was often selected for the more expensive items.

Although now having little relevance to modern-day farming, the nuts of the beech, known as mast, were, like acorns, used as fodder for the fattening of swine in the forests – the ancient right of pannage. The New Forest in southern England is one of the few places today where farmers still regularly turn their pigs out to forage each autumn. The nuts were also sometimes fed to poultry. Beechnuts are a good source of protein, minerals and trace elements, including potassium, magnesium and calcium, but they have never been much used for human consumption in Britain. However, there is some evidence to show that beechnut oil was used for cooking in France in the nineteenth century. Beech leaves used to be used for bedding and John Evelyn records that 'they afford the best and easiest mattresses in the world to lay under our quilts instead of straw; because, besides their tenderness and loose lying together, they continue sweet for seven or eight years, long before which time straw becomes musty and hard ... I have sometimes lain on them to my great refreshment.'

MYTHS, MEDICINE AND MUSHROOMS

Beech has long been a symbol of prosperity and it is rumoured that if a beech stick is carved with your greatest wish then it will come true. It was always known as the lovers' tree or the trysting tree, mainly because of its perfect 'notebook' bark, the ideal medium for incised romantic graffiti. The pendulous boughs of huge, spreading trees also provided perfectly private green bowers to shield many a secret rendezvous.

The beech would appear to have few useful medicinal properties. Nicholas Culpeper recommended the leaves as a poultice for 'hot swellings' and claimed that water collected from the hollows at the base of the tree would 'cure both man and beast of any scurf, or running tetters, if they be washed therewith'. ('Tetter' is a term applied to various skin diseases, such as ringworm, eczema or impetigo.)

Southern beechwoods are deemed the best fungal habitat in Britain. The rich humus beneath the trees encourages many different fungi and many of these will unite with the roots of the beech in a mycorrhizal relationship, each helping the other to obtain nutrients. This is how truffles survive. Some fungi are less considerate to their host, however, for they sound the trees' death knell. These killers include the massive bracket fungi, such as *Meripilus giganteus*, which may have caps up to 19 inches wide, and *Ganoderma adspersum*, which grows in great tiers as long as 39 inches across.

The last rays of the evening sun filter through beeches on the shores of the Lake of Menteith in Stirlingshire.

An ancient yew deep in the woods at Newlands Corner in Surrey displays the bulbous bole so typical of many of the yews in this very special woodland.

Yew

No other single tree species in Britain inspires such wonder or stimulates more debate than the mighty yew. The sheer enormity and antiquity of the most ancient specimens, which are usually to be found within the confines of churchyards, simply puts all other trees in the shade. The intrigue with yew is born out of its secretive nature.

Old emperor yew, fantastic sire,
Girt with thy guard of dotard kings,
What ages hast thou seen retire
Into the dusk of alien things?
What mighty news hath stormed thy shade,
Of armies perished, realms unmade?

FROM A POEM BY WILLIAM WATSON (1858–1935)

Yew berries – the bright pink aril conceals the poisonous kernel.

Nobody fully understands the yew's growth mode – this is a tree that sometimes drops into virtual stasis for several decades. It may push up new stems, drop aerial roots within a hollowed trunk or cause its outer limbs to arc to the ground where they layer and regenerate. It will thrive in the hollows of other trees or grow from seemingly solid rock. The oldest trees have lost the evidence of their true age, as their heartwood has long since rotted away, leaving cavernous hollows within. The symbolism of the yew is at once at variance with itself – a poisonous and gloom-laden tree, redolent of death yet containing also a promise of renewal and immortality.

The common or English yew (*Taxus baccata*) is most clearly defined by its sombre dark blue-green fronds, its broad yet conical form as an open-grown specimen, the poker-dot peppering of vivid pink berries in the autumn and the fine-fluted form of its bole, rich in hues of pink and mauve and various shades of red, green and brown, accentuated when the tree is wet. Yews will tolerate most types of soil but the presence of large numbers of veteran trees and woodlands indicates a distinct natural range across Britain, governed largely by their preference for chalk or limestone.

THE YEW IN THE LANDSCAPE

Yew favours that swathe of southern England from Kent to Hampshire (where the tree is so common that is has been dubbed the 'Hampshire Weed'). It is typically found on the steep combe slopes of the South Downs, where it frequently forms dense stands of woodland in which it is the dominant species and where little ground flora will grow beneath its dark canopy. The most evocative of all these woods is to be found at Kingley Vale near Lavant in Sussex. This is generally considered to be the finest yew woodland in Europe and it was considered special enough to be one of the first designated National Nature Reserves in 1952. From a distance the trees form a dark blanket across the hillside, with the occasional springtime relief of the creamy-green foliage of whitebeam. There have been yews at Kingley for at least 2000 years, it is said, but most of those growing there now are relatively young (less than 100 years old). However, at the wood's heart, in Kingley Bottom, lies a moody, brooding grove of bent and bowed antiquarians that are estimated to be around 500 years old. To walk among these trees, particularly at dawn or dusk, or on a murky fog-bound lonesome day, is to be overcome by the irrefutable aura of these other-world monstrous beings. Their bulbous, burred boles squat resolutely on the barren brown earth, arachnoid limbs clawing out from beneath the canopy to seek whatever support they may find. Dark rainwater runs like some lifeblood loss weeping from many a fork and crevice and, as the wind rises, the old ones creak and moan to one another of centuries long departed. Author, naturalist and ornithologist, W. H. Hudson, writing at the beginning of the twentieth century, echoes these sentiments in his book Nature in Downland: 'dark religious trees, with trunks like huge rudely-fashioned pillars of red and purple ironstone. One has here the sensation of being in a vast cathedral: not like that of Chichester, but older and infinitely vaster, fuller of light and gloom and mystery, and more wonderful in its associations.'

This mighty yew in Borrowdale was one of the famous 'fraternal Four' referred to by William Wordsworth in his poem Yew Trees, *written in 1803. Sadly, it was blown down in 2005, but the remnant stump is showing encouraging signs of regeneration, so with any luck the celebrated tree will live on.*

Yew takes its turn as an understorey species in many woods, often in association with ash or beech in southern counties, and seems more than happy in the deep shade that such trees will cast. The species is present in large numbers in the woodlands of the Wye valley, extending westwards to the Welsh borders. Here and there, where space and light allow, large yews have evolved, often being left on boundary banks or woodland edges as markers. In Shropshire there are some impressive hedgerow yew trees, once again providing excellent points of reference in the landscape. Perhaps these were once woodland trees, now set alone after the wood was cleared long ago, or maybe they are former boundary markers. No one knows for certain. Particular landmark yews are mentioned in various Anglo-Saxon charters and once a boundary had been established there was every reason for a landowner to keep a marker in place, replanting where necessary.

HOW OLD ARE YEWS?

Understanding what makes the yew's place in our natural and cultural history so special is an ancient story with many twists and turns. It is a common claim that the oldest living thing in Britain is a yew tree. However, there is a challenge to this view in the shape of certain ancient broadleaf coppice stools (particularly native limes) that naturalists believe may derive from trees of a similar vintage. To find the definitive answer to this conundrum is a quest that has occupied some of the world's finest botanists and archaeologists for more than two centuries.

In the early nineteenth century Alphonse de Candolle, the renowned Swiss botanist, was the first person to correlate ring-counts and girth measurements to estimate the age of a yew tree. After surveying and measuring many trees he concluded that an average growth of 1 foot of trunk diameter contained 144 annual rings – that is an annual growth rate of 0.04 inches radius width per year. Once this theory was launched it didn't take long for other naturalists to join the fray. Notable among the contenders was John Lowe, author of *Yew Trees of Great Britain and Ireland.* Published in 1897, this was the first authoritative work dedicated to the yew. Lowe considered that a mere 60–70 annual rings per foot was closer to the mark. Perhaps they were both right, for they had obviously sampled from a different selection of trees. Many different theories and estimates entered the debate throughout the twentieth century but it wasn't until about 1980 that research took a leap forward.

Over the last 25 years some new methods of assessing tree dates have been proposed. A couple of wood samples from the insides of ancient hollowed yews have been submitted to radiocarbon dating, the results in this instance showing that the trees were several hundred years old. However, since almost all heartwood of ancient yews has either disappeared or is in decay, it is not safe to rely on this technique because the samples might have come from more recent stem or root regrowth. Another method is to take core samples from the tree in order to correlate the annual rings with documented measurements, though the invasive nature of this process is a cause for concern.

The patterns of annual rings can also be helpful, since dendrochronology sets up templates of the growth patterns of individual species over several centuries, calibrated off samples of a known age. Once established, this pattern can then be cross-referenced with samples from other sites, enabling time-line comparisons leading to dating. This has been done in Britain with oak for the last 30 years but the yew model is relatively recent. Further research using the yew model should eventually be able to track back thousands rather than hundreds of years. The bristlecone-pine model in America, where dendrochronology first began in the 1920s, currently maps 9000 years.

There are other factors to consider too, since regional variations in climate, exposure and soil types can all have a dramatic effect on growth rates. An open-grown floodplain tree, or one in a sheltered valley in a warmer climate, is likely to grow much faster, with larger increments of annual growth, than one that struggles for a foothold on some storm-blasted northern crag. The fact that so many old yews are hollow, have sections missing and have developed buttressed protuberances merely adds to the problem. Bring into the equation those unexplained, occasional periods of growth stasis peculiar to yew and then multiply these factors over many centuries, and the outcome becomes increasingly uncertain.

Ultimately, most casual observers of yew trees are satisfied with a rough age guide and, frankly, anything in excess of 1000 years is pretty mind-boggling. Robert Bevan-Jones provides a useful table in his excellent monograph *The Ancient Yew*, published in 2002, though he insists that these are only rough estimates as there are so many different factors that can limit, accelerate or metamorphose the forms of yew.

3 feet girth	80–100 years old
10 feet girth	250–350 years old
13 feet girth	350–500 years old
16 feet girth	600–750 years old
20 feet girth	700–1000 years old
23–33 feet girth	1200–2000 years old

THE ANCIENT YEW

Many ancient yews have achieved national celebrity status due to an increased awareness of heritage trees as some of the most culturally and historically significant elements of the British landscape. Several organizations, such as The Tree Council, The Woodland Trust, The Ancient Tree Forum and the Ancient Yew Group, have campaigned for certain of these trees to be awarded Green Monument status, giving them the same level of legal protection accorded to other scheduled ancient monuments. Working with The Church of England, The Conservation Foundation gathered cuttings from a wide selection of ancient yews in the 1990s and distributed over 7,000 propagated Millennium Yews to be planted out in churchyards.

The Aberglasney Yew Tunnel in Carmarthenshire makes a unique garden feature. Thought to be around 250 years old, this row of trees were once bent over to form natural arches and where they met the ground they took root. This remarkable garden and its yews was all but lost until rediscovery and restoration during the 1990s.

The most famous of all of Britain's ancient yews is the shattered remnant of the Fortingall Yew in Perthshire. The Welsh naturalist Thomas Pennant visited this tree in 1769 during his first tour of Scotland, when he recorded its circumference as 56 feet 6 inches. When Judge Daines Barrington took measurements in the same year he recorded it as 52 feet and his is the figure that has always been accepted as the more accurate one. Variations in tree girths are best explained by the fact that some measurements were taken at chest height whilst others would have been at ground level. The earliest-known image of this particular yew is a sketch by Jacob George Strutt, which he published in his *Sylva Britannica*. This clearly shows a hollow shell of a tree with a funeral procession passing through the middle – a custom thought to guarantee safe passage to the afterlife and immortality. Dr Patrick Neill, who visited the tree in 1833, wrote in the Edinburgh New Philosophical Journal that:

> considerable spoliations have evidently been committed on the tree since 1769; large arms have been removed, and masses of the trunk itself carried off by the country-people, with the view of forming *quechs*, or drinking cups, and other relics which visitors were in the habit of demanding. What still exists of the trunk now [1833] presents the appearance of a semicircular wall, exclusive of the remains of some decayed portions of it which scarcely rise above ground.

These comments must have galvanized the local populace because a stone wall was later built to protect the tree from those who would plunder its timber. This wall, with its wrought-iron railings, still surrounds the tree today. A pegged circle, taking in the existing fragments of growth, indicates its maximum, original girth. Age

The Fortingall Yew in Perthshire, an engraving from The Forest Trees of Britain *by the Revd C.A. Johns, 1849, depicts a burial procession passing through the hollow heart of the tree, a ritual that was supposed to confer immortality of the soul upon the departed.*

estimates vary widely here, from the eminently plausible 3000–5000 years up to an improbable 9000 years old. Whatever it may be, it is almost certainly the oldest living plant in Britain, if not in the whole of Europe.

The very largest yews in Britain number about 40 specimens, with girths of 30 feet or more, which would make them around 2000 years old. Although there are many remarkable individuals in Kent, Surrey, Sussex and Hampshire, with outliers in places such as Tisbury in Wiltshire, Kenn in Devon and Ashbrittle in Somerset, roughly half of all the biggest trees are to be found in Wales and the neighbouring counties of Herefordshire and Shropshire. What makes all these yews so astounding, other than their sheer scale, is their breathtaking beauty and variety of form. Some of these ancients have acquired great notoriety. Legend has it that the Ankerwycke Yew at Runnymede in Berkshire bore witness to the signing of the Magna Carta by King John in 1215. This tree was certainly large enough to have been a significant landmark at the time. King Henry VIII is also reputed to have met with Anne Boleyn, who lived at nearby Staines, beneath the tree in the 1530s.

The hollows of ancient yews have given rise to many anecdotes over the years. In 1843 an account of a yew in Tisbury (then in Dorsetshire, now in Wiltshire) in *The Penny Magazine* described how 'a few years ago a party of seventeen persons breakfasted within its capacious bole'. The tree then had a girth of 37 feet. A few years later someone decided to fill the hollow with concrete. Such vandalism would not be allowed today but the tree has survived nevertheless. Recent measurements put its girth at 31 feet. Another hollow yew, in Crowhurst in Surrey, still has its hollow bole protected on one side by a door that was built to enclose a small 'tree room', which was fitted out with a table and chairs. A dispossessed family of nine were reported to have lived in it for a while some time during the nineteenth century. The splendid Much Marcle Yew in a Herefordshire churchyard has a fine hollow, complete with bench. It is a favourite venue for local wedding photographers, who offer it as a memorable setting for the couples who have married in the adjacent Hellens and St Bartholomew's Church.

Some of the massive yews that were particularly well documented in the nineteenth century have been lost. One of the most famous of all, the Selborne Yew in Hampshire, was originally described by The Revd Gilbert White in his *Natural History and Antiquities of Selborne* of 1789. White notes that he measured the male tree at 23 feet in girth. Some 200 years later Britain's most prolific tree recorder in recent times, Alan Mitchell, found this to have increased to almost 26 feet. Although not one of Britain's very oldest yews, this one still comes in at about 1200–1400 years old. In January 1990 violent storms knocked over the tree. The villagers tried to pull it upright with heavy lifting gear but the root system had obviously been badly damaged and it died soon after. Its stump remains in the churchyard as a memorial to White and another yew grown from a cutting has been planted near by to ensure genetic continuity.

There is a fascinating cross-reference concerning a once famous yew that used to grow in the churchyard in Dibden in the New Forest. Prideaux John Selby describes it in his book *A History of British Forest Trees*, which was published in 1842. If he had only read the solemn report that appeared in The Saturday Magazine on 4 February 1837 he would have known that this tree had been destroyed in a hurricane in 1836.

THE SPIRITUAL YEW

Most of Britain's great yew trees stand in close proximity to churches and this throws up a whole volley of new questions. Why is the tree there? Was it there before the church and, if so, what was its significance? It is likely that many of these churchyard yews are much older than the buildings, especially as churches were often built on existing pre-Christian sites. When Pope Gregory dispatched St Augustine from Rome in the seventh century to convert the pagan Britons, he clearly briefed him not to destroy any places of worship that he found but to convert them into Christian churches. The words 'kirk' and 'church' probably come from the Celtic *cerrig*, meaning a 'stone' or 'circle of stones'. It is conceivable, therefore, that the first Christian churches in Britain were established in the groves sacred to the Druids, who performed their rites within a circular stone enclosure. The circle was an important symbol for the pre-Christian Celts. Reinforced by the sun and moon (and, had they but known it then, their own planet) the circle signified continuity and renewal and because it had no corners there was no place for the black spirits to hide.

The presence of a circle of yews planted within a circular churchyard points strongly to the pre-Christian belief in religious geometry, though there is no hard evidence of this. About 25 examples of tree circles have been recorded, many in Wales, and Bevan-Jones notes that some of them are now incomplete. There is quite a wide variation in the age of yews that remain, some may be as much as 1500 years old, which would mean they existed before the first Christian missionaries arrived in Britain in the sixth century.

Improved record-keeping over the last 500 years allows us to be more certain about when a yew tree was planted in a churchyard, though it still doesn't answer the question why. Photographer and writer on matters spiritual and mystical in the natural world Michael Jordan addresses this issue in his book *The Green Mantle* and attempts to understand the yew's mystique:

> To some extent the dearth of information [in regard to sanctification of plants] is explained by the fact that cult – Christian, pagan or otherwise – makes a timeless demand for secrecy. Esoteric cult and ritual are reliant to a large extent on fear of the unknown. Fear has always been, and continues to be, the 'muscle' of religion that persuades its congregation to maintain the letter of the faith. Fear, therefore, amounted to the 'powerhouse' of the earliest priests and shamans and the secrecy that generated the fear was the instrument through which the few held sway over the rest. Bringing cult into the public eye put the power at risk of being dispelled in the presence of familiarity.

The incredible leaning yew in the churchyard at Bedhampton in Hampshire.

The dense nature of the tree was also thought to form a good windbreak, thus protecting the church from storms. Yew was often used as a substitute for palms on Palm Sunday, both for decorating the church and to be given to the congregation, mainly because Britain had no palms. Sometimes the yew sprigs would be burnt and used for the following year's Ash Wednesday ceremonies. Churchwarden accounts record that box, laurel, broom and goat willow were all used for this purpose as well, which would indicate that the yew had no particular significance but was merely easy to come by.

In the absence of hard evidence, religious traditions and folklore have developed their own theories, for example the Christian belief that the white sapwood and red heartwood of the yew represented the body and blood of Christ. The yew's evergreen foliage and undisputed longevity make it an ideal symbol of immortality. It was commonly believed that a yew would drive away evil spirits, thus protecting the church and all those buried thereabouts. The tradition of strewing yew sprigs in a grave or placing them within coffins was probably borrowed from the Romans. Greek and Roman funeral pyres were often fuelled by yew and other evergreens.

In some areas of Wales to say that someone was 'sleeping under the yew' merely meant that they had died but in other parts of the country it was generally considered to be a risky business to lie or sleep beneath a yew, as one might be dragged into the netherworld by the dead buried below or perhaps set upon by the witches, ghosts and demons hiding within the tree. Shakespeare draws on this superstition in Macbeth, where the witches on the blasted heath stir their bitter cauldron with

Liver of blaspheming Jew, Gall of goat, and slips of yew Sliver'd in the moon's eclipse.

In 'The Grave', the eighteenth-century Scottish poet Robert Blair paints an extremely gloomy portrait of the churchyard yew:

Well do I know thee by thy trusty Yew,
Cheerless, unsociable plant, that loves to dwell,
'Midst skulls and coffins, epitaphs and worms;
Where light-heeled ghosts, and visionary shades,
Beneath the wan cold moon (as fame reports),
Embody'd thick perform their mystic rounds
No other merriment, dull tree is thine.

THE USEFUL YEW

There is perhaps a more prosaic explanation for the occurrence of yews in churchyards and some theories point to the tree's poisonous nature. It is particularly harmful to cattle and horses, the toxins becoming more concentrated when the foliage is desiccated, so maybe this was a way of ensuring that farmers kept their stock out of the churchyards. Strangely though, sheep, goats and deer all seem to be able to browse the tree without deleterious effects. The seed inside the bright pink fruit is also poisonous, although the sweet, fleshy surround – known as the aril – is perfectly palatable. Birds frequently eat these, but usually regurgitate the seeds.

Because most parts of the yew are toxic to human beings the tree has been used sparingly for medicinal purposes, although there are records of preparations made to treat epilepsy, nausea, dizziness, Ménière's disease, rheumatism, arthritis and liver and urinary problems. However, in recent years, yew has offered a ray of hope to cancer sufferers. Researchers in America in 1962 discovered that an extract from the bark of the Pacific yew (*Taxus brevifolia*) contained anti-cancer properties, which would later

This representation of the Battle of Cressy (Crecy) from The Saturday Magazine *of 1836 clearly shows the English archers with their superior longbows inflicting their worst upon the Frenchmen who were armed with crossbows, which took much longer to rearm and fire.*

be identified as Taxol – a complex polyoxygenated diterpene. The problem was that massive quantities of yew clippings were required to make relatively small amounts of the drug. Common yew was found to contain a closely related compound to Taxol, called 10-deacetyl baccatin III, which could be chemically modified into a semi-synthetic version of Taxol called Paclitaxel. This was first made available to the public in 1995 and has subsequently been used in conjunction with other treatments for certain types of ovarian, breast and lung cancer, with very encouraging rates of success. Many large estates and gardens are now carefully harvesting their yew clippings and sending them off for processing.

Yew-wood is arguably the most beautiful of British native hardwoods, being richly coloured in reds, oranges, yellows, cream and occasionally mauve or purple. Because the yew is so wayward and unpredictable in its growth pattern, often being flawed by knots and voids, it is extremely difficult to work and cabinet-makers often reserve it for the fine detail or inlay in larger pieces of furniture. However, when a substantial piece is made in yew it is a beautiful sight to behold and the extreme hardness and smooth finish is a tactile thrill.

Yew has long been renowned as the provider of staves to make longbows and there is no doubt that for several thousands of years this was the best weapon for both warfare and hunting. Archaeological evidence of yew bows is fragmented. There are a few Neolithic and Bronze Age examples and then, with the odd exception, there is nothing on record until the recent raising of the Tudor warship the Mary Rose, when many yew bows were recovered. By strange coincidence, the English scholar Roger Ascham, who published his treatise on archery, *Toxophilus*, the year before the Mary Rose sank in the Solent in 1545, noted: 'As for brasell, elm, wych, and ashe, experience doth prove them to be mean for bowes; and so to conclude, ewe, of all other things, is that whereof perfite shootinge would have a bowe made.' He went on to discuss how a bowyer should select the wood: 'every bowe is made of the boughe, the plante or the boole. The boughe is knotty and full of pruines; the plante is quick enough of caste, but is apt to break; the boole is the best.' And he also gave advice on choosing the best bow: 'If you come into a shoppe and fynde a bowe that is small, longe, heavye, stronge, lyinge streighte, not wyndynge, nor marred with knottes, gaule, wyndshake, wem, freat, or pinch, bye that bowe on my warrant.'

It was generally accepted, even in the sixteenth century, that the best wood for bows came from Spain and Italy, so yew-staves were imported in large quantities. There are a few churchwarden accounts of yews being cut for bows: in Ashburton in Devon, for example, there is a record from 1558 that refers to 'lopping the yew-tree' and it gives details of the payments made to the bowyer for his work. If yews had been regularly harvested for this purpose, then surely there would be more evidence of coppiced trees in churchyards or perhaps considerably fewer trees altogether.

YEW AND THE ART OF TOPIARY

The training and sculpting of yew trees in gardens has been periodically fashionable in Britain for the last 300 years. The earliest evidence comes from the first century, when the Roman writer Pliny the Younger described his Tuscan garden, but box was then the chosen tree for most of the intricate designs. There are brief accounts of topiary observed at Hampton Court Palace at the end of the sixteenth century but the practice did not become widespread until the end of the seventeenth century, when it was introduced into Britain from Holland during the reign of William and Mary.

John Evelyn was the first actively to encourage the use of yew for hedges and topiary in his *Sylva* of 1664 and the next 50 years would see English gardens set with all manner of decorative yew trees, arbours and walks. Tragically, most of this was destined to be swept away by the Landscape Movement (William Kent, Charles Bridgeman, Capability Brown and others) later in the eighteenth century in what gardening writer Ethne Clarke adroitly terms 'one of those collective acts of cultural vandalism that is difficult to forgive'.

Thankfully, one fine example of a late seventeenth-century topiary garden has survived at Levens Hall in Cumbria. Designed by Guillaume Beaumont, the gardens were laid out in 1694 for the owner Colonel James Grahme, who had consulted his friend

Levens Hall in Cumbria is undoubtedly the finest topiary garden in Britain, the great Umbrella Tree making a dramatic centrepiece. The impressive cedar of Lebanon in front of the house was blown down in a gale in 2005.

Evelyn about his plans. The Umbrella Tree, the largest yew and the centrepiece of the garden, is thought to be 400 years old, which would indicate that it was either an existing specimen that had been topiarized or one that had been transplanted into the garden as a mature tree. The project to complete the garden took 30 years and William Gilpin, the eighteenth-century writer, educationalist and instigator of 'the picturesque' in landscape appreciation, described it as 'all that is best in Landscape and Design'. That it has survived intact is a miracle and probably all due to the fact that the property has been continuously owned by the same family ever since. The variety of forms in this garden is breathtaking. As well as the usual geometric shapes, such as bells, spirals, pyramids and chess pieces, there are many Levens specialities: Queen Elizabeth and her Maids of Honour, Dutch Oven, Judge's Wig, Bellingham Lion, Wigwam and, of course, Umbrella. Relief in colour and texture is provided by golden yew and box, early nineteenth-century additions.

The fashion for topiary in Britain has waxed and waned several times since Levens was laid out, with two main revivals in interest in the late eighteenth and late nineteenth century. Even so, the latter revival had its fervent protagonists and detractors. Some saw topiarizing as an art but others considered it to be the unnaturalistic and abhorrent containment of nature's true forms. Whilst topiary might once have been the medium of the gentry's fine gardens, where hordes of gardeners were employed to keep everything trimmed immaculately, the nineteenth century began to see it creeping into the vernacular. Garden snobbery and one-upmanship seems to have been a feature of Victorian life and if it were possible to fashion an element glimpsed in some grand garden in order to steal a march on one's neighbours, then a piece of topiary became that ideal statement. The legacy today is a countrywide collection of very individual and occasionally extremely bizarre and amusing topiary gems, many of which suddenly appear as a pleasant surprise to the passer-by of cottage garden or urban villa.

Yew was also a popular choice for hedges, its dense nature a perfect protective windbreak and its dark foliage a splendid foil for all manner of flowering plants and broadleaf trees with decorative foliage. A particularly massive tumbling example can be found in Powis Castle gardens near Welshpool. Creating something of a natural fortress is the tallest yew hedge in the world, in front of the Bathurst Estate in Cirencester, Gloucestershire. It's 300 years old, 40 feet high and 150 feet wide and takes two men, using a cherry-picker, twelve days to trim. The garden designers' extension of the yew-hedge theme was the construction of yew mazes.

The huge yew hedge in the gardens of Powis Castle, near Welshpool.

Topiary was certainly not the prerogative of the gentry as this wonderful Edwardian postcard of Yew Tree Cottage near Pontypool demonstrates.

Originally an Elizabethan innovation, one of the oldest-surviving examples is at Hampton Court Palace, dating back to 1690. Mazes had something of a revival in the early nineteenth century and have continued to fascinate ever since. The late Felix Dennis, publisher, poet and bon viveur, planted his OZ maze of yew on his estate in Warwickshire twenty-five years ago, and serves as part of his wonderful (and humorous) tree legacy (See also Hedges and Hedgerow Trees).

THE YEWS OF FLORENCE COURT AND DOVASTON

Worthy of mention as an adjunct to the common yew are the two sports (or mutations) of the tree that have claimed their own place in history. A fastigiate or upright form of the tree was discovered by chance in the Cuilcagh Mountains above the eighteenth-century house and estate of Florence Court in County Fermanagh in around 1770 by George Willis, who was a tenant farmer of the estate's owner, the Earl of Enniskillen. Two specimens were taken from the mountains. Willis planted one of these on his farm, where it survived until 1865. The other tree is still living in the grounds of Florence Court. In the early nineteenth century horticulturists began to take an interest in this tree, recognizing its potential as an ideal subject for topiary. Cuttings were made available commercially after 1820 and this one yew has been the mother tree of all the Irish yews in the country, if not the world. It became a popular choice for churchyards throughout the nineteenth and twentieth centuries.

By strange coincidence, at almost the same time (1777 to be exact), a Shropshire landowner, John Dovaston, purchased a yew sapling from a local pedlar in West Felton. He was having trouble stabilizing the earth around a newly dug well on his estate and thought the tree's fibrous root system would solve the problem. It did, and by way of a bonus the sapling grew with a beautiful weeping or pendulous habit. Moreover, it was discovered that this remarkable specimen bore both male and female flowers (unusual for yew, which is usually dioecious, that is having separate male and female trees). Careful cross-pollination between both types of flower produced trees that were true to the weeping form, thus ensuring its future. The original Dovaston Yew survives today within a modern housing development, where it is cherished by the village community.

The Brahan Elm in Easter Ross is currently the champion wych elm in Britain with a girth of 23 feet and towering 84 feet in height.

Elm

There was a time when the elm rivalled the oak as the landmark tree of lowland Britain. The oak may be a treasured national symbol but it is the distinctive silhouette of a winter elm against a storm-ravaged sky or billowing tiers of emerald spring foliage that older generations remember with nostalgia. The elms housed rookeries and stood sentinel along the hedgerows. Constable captured their dignified beauty on canvas and poet John Clare celebrated them in his poem 'The Fallen Elm'. Then, the landscape changed for ever, with the arrival of Dutch elm disease.

Oh, well-aday for you, our quiet friends!
If you should perish, what should make amends?
England and Elms – the two words run together
Like sun and happy thoughts in summer weather,
When one without the other cannot be.
What child of England, driven to foreign realms,
Brooding on home, but pictures English elms
In the green meadows, or along the lane
The homeward path he yearns to tread again?
Or sees their branches 'gainst the winter sky
Snow-laden, woven in delicate tracery?

FROM 'AN ELEGY ON ELMS' BY A. RUTH FRY (1878–1962)

The winged seeds (samaras) and new leaves of wych elm.

There are three principal elms to be found across Britain – the English elm (*Ulmus minor* var. *vulgaris*), the smooth-leaved elm (*Ulmus minor* subsp. *minor*) and the wych elm (*Ulmus glabra*). Identification is a complex matter because there are so many hybrids, varieties and clonal variations, particularly of the smooth-leaved elm. However, there are some common features, such as leaf shape, that can help to distinguish the elms in general from other trees.

In spring the elms tend to be one of the earliest trees to flower, usually in late February or early March. These flowers, which appear as small purplish or pinkish tufts at the ends of the twigs, are bell-shaped perianths (the outer part of the flower), each containing four to nine (usually five) stamens. Their colour comes from the anthers (the part of the stamen containing the pollen) and they have no petals or sepals. The vivid green winged seeds (called samaras) form before the leaves emerge; sometimes a seed-laden tree can give the impression from a distance of already being in leaf.

Seeds of the wych elm are often fertile and this is usually their main method of reproduction. However, seeds of other elms are almost always infertile and so they tend to reproduce by suckering. Many of the hybrids and varieties to be found today are clonal descendants of trees that were growing when the British climate was warmer, which means that all elms were once fertile enough to hybridize freely. Leaves tend to emerge in late April or early

May. They are invariably asymmetrical in shape, with a stepped or uneven base where they meet the leaf stem, have serrated edges and are pointed at the tip. Elm leaves can vary dramatically in shape, but those of the English elm and the wych elm are always rough to the touch on their upper surface and downy on the underside, with a smaller width-to-length ratio than the smooth-leaved elms. As one would expect, the leaves of the smooth-leaved elm are smooth.

THE ENGLISH ELM

The English elm (*Ulmus minor* var. *vulgaris*), once thought to be a species in its own right (*Ulmus procera*), is now considered to be a subspecies of the smooth-leaved elm. Exactly where the tree originated is uncertain but it is probably an introduction from continental Europe. Dendrologist and woodland expert John White suggests that it closely resembles elm trees native to northern Spain and that migrating Neolithic peoples might have brought them to Britain, most likely as a source of green fodder in late summer when grass was exhausted. Indeed, in 1911 the Irish plant collector Augustine Henry noted that elms in the Royal Park in Aranjuez bore a striking similarity to English elms, but John Evelyn states in his *Sylva* that these trees were taken over by Philip II from England in 1575. Henry was doubtful that the trees he saw were in excess of 300 years old and he considered them to be of a much later planting date. However, it is undeniable that the tree has been widespread throughout the lowlands of Spain, although whether or not it is a true native or arrived there from another country is unknown.

Over thousands of years the English elm became widely established throughout the Midlands and the southern counties of Britain, usually spreading via human intervention or by reproducing naturally from suckers. The tree's characteristic profile – tall,

Handsome smooth-leaved elm in springtime near the Cambridgeshire village of Whaddon.

The two greatest English elms in Britain, both a little over 20 feet in girth, grow in Preston Park in Brighton. Sadly, the rear tree in this photograph succumbed to Dutch elm disease in 2019 and had to be felled.

straight and billowing, broadening at the crown – is the archetypal elm beloved of artists and poets. It is the unmistakable shape that was so sorely missed from the lowland landscape as the 1970s progressed. This was the tree that inspired the artist, writer and naturalist Gerald Wilkinson to write his excellent monograph and homage to the elms, *Epitaph for the Elm*, in 1978. He was the first person to write an accessible overview of the British elms, exploring their cultural history and significance, and he makes a valiant attempt to unravel their botanical interrelationships. Wilkinson lived in Oxfordshire, which was one of the foremost regions for the English elm. His description of their demise is succinct yet poignant:

> The elms are dying. Gaps appear in familiar lines of trees that we never bothered to think of as elms. Half the tall trees, and many smaller ones, in the roadside hedges seem to have

> been elms, we notice, now they are so unhappily conspicuous ... Great elms, landmarks to nowhere in particular, are shattered from green to grey, and will soon be gone, never to be replaced in our lifetime ... Little woods here and there which seemed to be of no particular tree now sadly, pointedly, show what they were. Houses and cottages once sequestered among two or three friendly trees are now threatened by their dead shapes ... We took them for granted as part of our characteristic lowland landscape.

THE SMOOTH-LEAVED ELMS

The smooth-leaved elms (*Ulmus minor* subsp. *minor*), sometimes referred to as small-leaved, narrow-leaved or field elms, were once also thought to have been introduced from Europe, perhaps by the Romans. However, pollen analysis has recently shown that these elms were here before the Neolithic settlers arrived, so it is more than likely that the trees colonized Britain in the same way as many other species returning in the post-glacial period. As the name indicates, this is an elm with a distinctive smooth leaf, which is often quite shiny on the upper surface; it has a much more elongated shape than the English elm leaf. This is the elm of East Anglia, a tree that has avoided Dutch elm disease in several localized concentrations rather better than many of its cousins, although none of them is totally immune. Two particular colonies, around Boxworth in Cambridgeshire and on the Dengie peninsula in Essex, have fared better than most and many impressive large elms are still in evidence here. All these very individual colonies of elms, and there could be hundreds of them, are clonal, as they have all grown from suckers. For example, the Cambridge botanist Oliver Rackham identified 29 distinct elm clones on a 40-acre site at Buff Wood in Cambridgeshire. It may have been their identical DNA that hastened their demise in the face of Dutch elm disease but similarly it could have been the isolated situation of some colonies that saved them. Like most elms, this tree has changed its Latin name in the light of ongoing research and different ways of thinking. Prior to 1946 it was identified as *Ulmus nitens* and more recently as *Ulmus carpinifolia*.

A handful of varieties and cultivars of smooth-leaved elm are worth noting. Cornish elm (*Ulmus minor* var. *cornubiensis*) can be traced back as far as 1633, when it was first identified by Hampshire botanist John Goodyer. In maturity this tree has a very distinctive profile – tall and relatively thin with many epicormic shoots all the way up the trunk, giving it a somewhat ragged appearance. Large trees have disappeared from the Cornish scene but it still survives in hedges and thickets. Lock's elm or Plot's elm (*Ulmus minor* var.

The windblown form of an exposed Davey elm in a Cornish hedgerow. Ulmus x hollandica *'Daveyi' is restricted to Cornwall and is considered a hybrid of wych elm and Cornish elm.*

lockii) is again a rather tall, slender tree but without the multitude of side shoots seen on Cornish elms. Supposedly first identified by Dr Plot near Banbury in Oxfordshire in 1677, the main stronghold of this tree was east Nottinghamshire and Lincolnshire. Land improvement, intensive farming and disease have all contributed to its decline in these regions now. The Guernsey elm, otherwise known as the Wheatley elm (*Ulmus minor Sarniensis*), is a fine conical tree, now wiped out in the Channel Islands by Dutch elm disease in spite of the best efforts of the local authorities to save it. An impressive row thrives in Preston Park in Brighton.

THE WYCH ELM

If there were ever any doubt about the native status of the other elms, then the wych elm (*Ulmus glabra*) has been commonly accepted as a native species by all authorities. Wych elm grows throughout Britain and, probably due to its wide distribution, often occurring in quite isolated locations, it has managed to overcome the sweep of Dutch elm disease. It is believed that its habit of not throwing suckers protected its root systems and it is evident that the bark beetles don't seem to favour it for feeding. This is the elm most prevalent in Wales, northwest England and Scotland (where it used to be known as the mountain or Scotch elm – *Ulmus montana*). This species accounts for some of the largest surviving elms in Britain. A champion specimen is the Brahan Elm in Easter Ross in the north of Scotland: it is a thumping great tree, 84 feet high, with a girth of 23 feet, and it is still in very good shape.

The leaves of the wych elm are very distinctive. They are the biggest of all the elms, sometimes growing to a length of 6 inches, and examination of a leaf taken from a principal lateral bough will usually show that the basal lobe on its longer side overlaps

This rendition of two weird old wych elms near Cradley, on the Herefordshire and Worcestershire border, taken from Woolhope Naturalists' Field Club journals of 1869, seem to be more burr than tree.

or conceals the leaf stem. A distinct horn or spur will sometimes develop on either side of the main leaf tip.

Where it grows in eastern England the wych elm is prone to hybridizing with the smooth-leaved elm. One of the most successful outcomes of this habit is the Huntingdon elm (*Ulmus* x *hollandica* 'Vegeta'), which, as the name suggests, first arose in a nursery in that part of Cambridgeshire, around 1760. Many respectable-sized trees are still in evidence across the Midlands and it has proved to be reasonably resistant to Dutch elm disease.

The word 'wych' has nothing whatsoever to do with witches or black magic, although this belief persists. Botanist Revd C. A. Johns in his book *The Forest Trees of Britain*, published in 1882, reports: 'In some of the midland counties the name seems to have originated the notion that it is a preservative against witchcraft, and a sprig is inserted into a hole in the churn by dairymaids, in order that the butter may come freely.'

J. C. Loudon ascribed 'wych' to a water spring, citing such places as Droitwich and other spas, and it is true that the tree does seem to favour sites near rivers and streams. Wilkinson reckons that 'wick' or 'wych' was much more likely to mean a farm or dairy farm (which correlates quite neatly with Johns's account). Botanist and social historian Geoffrey Grigson, on the other hand, believes the word meant 'switchy' or 'pliant', which would fit in with its traditional use for making longbows. The twelfth-century Welsh cleric Geraldus Cambrensis bears this out, for he describes his encounters with Welshmen whose bows 'are not made of horn, or ivory, or yew but of wild elm, and not beautifully formed or polished, quite the opposite: they are rough and humpy, but stout and strong nonetheless.' Since wych elm has always been widespread in Wales and is the only elm that cleaves well, this would appear to be an accurate observation.

THE USEFUL ELM

Elm timber has always been highly prized, second only in importance to oak as a building material. Because it tends not splinter, it has been used for beams, joists and floorboards, and for weatherboarding or cladding because it wears well in wet climates. This has also made it the timber of choice for ships' keels, mill wheels, lock gates, paddles and underwater piles. When London's old Waterloo Bridge was demolished in 1936, the elm piles upon which it had rested for 125 years were found to be in perfectly sound condition. The wood was chopped up and made into millions of souvenirs for people who wanted a piece of 'Waterloo elm'. The timber was frequently used in the construction of water pumps and underground water pipes. Elm pipes laid for the New River Scheme in London in 1613 were dug up in 1930 and they were still intact. The process of boring out these pipes with long augers must have been a laborious job but in the eighteenth century a mechanized method was invented whereby water wheels were connected both to the auger and a progressive racking forward of the elm logs. Elm was a most important timber for the old wheelwrights for making the naves or hubs of cartwheels. Broad boards of elm were always most sought after by coffin-makers, a tradition that has died with the diminishing elm stocks. Elm has long been chosen for the seats of Windsor chairs, as it works well with an adze without splitting – the craftsmen who did this were known as 'bottomers'. John Evelyn notes that shovelboard tables (shove-halfpenny boards) were often made of elm and these can still be found in some British pubs. Elm turns well and it is good for making bowls, a speciality being the intricate turning of nests of several bowls from a single block of wood.

This 1930s view of Elms Lane in Sudbury, Middlesex, shows the eponymous trees with their very distinctive profiles – the sort of treescape that has almost completely disappeared over the last fifty years.

Elm has always had many useful medicinal applications. One of the tree's vernacular names in Berwickshire was 'chew-bark' and people would sometimes chew a piece of the inner bark when they had a sore throat. In his *Complete Herbal* of 1653 Nicholas Culpeper claimed that: 'The leaves or the bark used with vinegar, cure scurf and leprosy very effectually' and, perhaps of interest to many people, even today, 'the roots of the Elm, boiled for a long time in water, and the fat arising on the top thereof, being clean skimmed off, and the place anointed there-with that is grown bald, and the hair fallen away, will quickly restore them again.'

In essence the elm has long been valued for its soothing and healing properties and could be taken internally or applied externally. Evelyn, who must have been familiar with Culpeper's advice, confirms this:

> The Green leaf of the Elm, contused, heals a green wound or cut, and, boiled with the bark, consolidates fractured bones. All the parts of this tree are abstersive [cleansing or purging], and therefore sovereign for the consolidating of wounds; they assuage the pains of the gout; and the bark, decocted in common water, to almost the consistency of a syrup, adding a third part of aqua vitae is a most admirable remedy for the ischiadica, or hip pain, the place being well rubbed and chafed by the fire.

THE ELM IN THE LANDSCAPE

Historic accounts and old engravings and photographs reveal that elms were once the most popular trees selected for formal avenues in Britain. Without exception, all these beautiful landscape features have now disappeared. Some avenues have been replanted, though usually with common limes rather than elms.

The fashion for planting avenues began after the Restoration of 1660. In 1680 one of the best-documented examples of this period was created: Charles II's famous Long Walk in Windsor Park. The king turned to Evelyn for advice and it was decided to plant a quadruple avenue of wych elms. It took about 1652 trees altogether and was 2¾ miles long. Many of the trees had disappeared by the late nineteenth century and in 1943 it was decided that because even more of them had since died of Dutch elm disease the avenue could not be saved. All the remaining wych elms were felled and replaced with London planes and horse chestnuts (a transition that had already begun in 1921).

Elms were generally popular with plantsmen and designers for they grew easily, transplanted successfully as quite large trees and were remarkably tolerant to a variety of difficult situations. They also resisted the effects of pollution, which made them suitable for the urban environment. In coastal areas they were perfectly at home and stood resolute against the salt-laden winds. Designers of the Landscape Movement of the eighteenth century may have despised the previous rigid and, as they saw it, artificial formality of Britain's gardens and parklands, but they were still able to appreciate the virtues of the elm as part of their contrived 'natural' landscapes. Even avenues were permitted, though winding in form and sympathetic to the natural contours.

In an anecdote entitled 'The Battle of the Elms', published in *After the Elm* in 1979, landscape and environmental writer Kathy Stansfield points out that 'there were times when the elm symbolized not only the delights of the landscape and the craftsman's labours, but the conflict between nature and technological advance.' Apparently there were ten fine elms in Hyde Park, to the south

One of the mature elms inside the newly erected Crystal Palace, as published in The Illustrated London News *of 25 January 1851.*

of the Serpentine, where it was proposed that the Crystal Palace should be built. In 1850 the rather grandly named Colonel Charles de Last Waldo Sibthorp MP mounted a campaign to save the trees. He was only partially successful, as by the time Joseph Paxton's iron and glass design was accepted some of the elms had already been axed. The Colonel and public pressure persisted, however, and Paxton was persuaded to amend his plan. He agreed to change his proposed flat roof into a domed one in order to accommodate three huge elms within the building. The irony is that nature had the last word: the imprisoned trees attracted a multitude of sparrows, that deposited their droppings on to the exhibits below.

BIRTH AND DEATH ATTENDS THE ELM

References to mighty old elms are common in nineteenth-century works and a few beautiful engravings exist of several of these craggy veterans. Like some ancient oaks, a few have become famed as gospel trees or hanging trees, whilst other hollow elms have provided hiding places for smugglers' booty or served as intimate dining rooms. In his *Sylva Britannica* Jacob George Strutt mentions what must have been one of Britain's biggest elms – the Crawley Elm in Sussex, which in 1830 was 70 feet high and 61 feet in girth at the base, with a hollow inside that was 35 feet around and closed with a door. Loudon records that 'a poor woman once gave birth to a child in the hollow tree'. A Mrs Smith, who lived opposite the tree in the early twentieth century, told Augustine Henry that she'd once seen 12 people sitting down for tea inside it.

Such intimacy with an elm is not always a good idea. English elm in particular has a reputation for suddenly shedding large and perfectly healthy-looking boughs for no obvious reason. Henry noted that this tended to happen in calm weather, often after rain.

The most likely explanation for this is that it is the tree's response to stress from various causes, such as drought, compaction or disease. It has been shown that at the point of the break the cells of the wood actually explode. This is perhaps the reason for the tree's popular name in some parts of the country: 'the widow maker'. Rudyard Kipling seems to have known about this and, in his poem 'A Tree Song', he wrote:

Ellum she hateth mankind, and waiteth
Till every gust be laid,
To drop a limb on the head of him,
That anyway trusts her shade.

DUTCH ELM DISEASE

When Ruth Fry wrote 'An Elegy on Elms' in the 1930s she was capturing a moment of history, for she was witnessing the effects of Dutch elm disease. It had arrived in Britain in 1927 and the first recorded casualty was on a golf course in Totteridge, Hertfordshire. The disease is Dutch by name only rather than origin: it was first recorded in France in 1918 and then Holland around 1920. This early epidemic, which was a less virulent strain than the one that swept across the country in the late 1960s, killed an estimated 10–20 per cent of the elm population. Some were partially affected but were strong enough to recover. The epidemic peaked in 1936 then, with exception of a few localized outbreaks, it subsided.

Similar pathogens had almost certainly wreaked havoc with elms prior to the twentieth century. Pollen records indicate that around 4000 BC the elm seemed to disappear in Britain for a long period. The reason might have been land clearance or unfavourable climatic changes but the possibility of disease should not be ruled out. Records show that elm disease occurred in France in 1818 with an outbreak towards the end of the nineteenth century in England, as one of Britain's most revered countryside writers, Richard Jefferies, confirmed in *Nature Near London*, published in 1883:

> There is something wrong with the elm trees. In the early part of the summer, not long after the leaves were fairly out upon them, here and there a branch appeared as if it had been touched with red-hot iron and burnt up, all the leaves withered and browned on the boughs. First one tree was thus affected, then another, then a third, til, looking round the fields, it seemed as if every fourth or fifth tree had thus been burnt ... Upon mentioning this I found that it had been noticed in elm avenues and groups a hundred miles distant, so that it is not a local circumstance.

The 1967 outbreak of Dutch elm disease was perhaps the most devastating that Britain had ever experienced. It entered the country via cargoes of rock-elm logs from Canada and it was no coincidence that the first trees to be affected were in the areas around the ports of London, Bristol and Southampton. Confirmation came in 1973 when a Customs officer at Southampton Docks found a consignment of rock-elm logs from Toronto that displayed stains in the wood typical of Dutch elm disease. Further investigation revealed beetles and their breeding galleries and laboratory tests later confirmed that the fungus in the logs was virtually identical to the one that was sweeping Britain. This time it was highly contagious and was lethal to almost all elms. The English elm was particularly susceptible. The epidemic eventually killed more than 60 million elms. Around 1990, after Dutch elm disease appeared to be waning, it suddenly took hold again and still shows no sign of abating.

The current epidemic of Dutch elm disease is caused by the microscopic fungus *Ophiostoma novo-ulmi*, introduced into the trees by two species of bark beetles – principally *Scolytus scolytus* but also *Scolytus multistriatus*. These beetles lay eggs in galleries, which they excavate through the underbark, next to the sapwood. When the larvae emerge they burrow through the sapwood, taking the spores of the fungus with them. The fungus multiplies and eventually blocks the water-conducting vessels of the tree. A twig broken from an infected tree and cut diagonally will reveal a brown ring in the infected xylem (water-transporting) tissue. First signs of infection

English elms in the very heart of Brighton. There are more elms in Brighton than any other single place in Britain.

are premature withering and yellowing of the leaves in summer. The tree then progressively dies back. The roots do not always die and will often send up new suckers, many of which may share the same root system, which will thrive for several years until the trees are perhaps 16–32 feet tall and have put on the rough bark. Unfortunately, this attracts the beetles once more and the whole cycle begins again. Alternatively, a tree's roots may fuse with those of neighbouring trees, thus passing on the disease by direct contact.

Dutch elm disease can be controlled by injecting the tree with fungicides or spraying it with insecticides but these are usually short-term measures. In the early 1970s, Rob Greenland, Brighton and Hove Council's arboriculturist, took stock of the situation and decided that a strict regime of pruning out infected boughs and felling infected trees the instant the disease was identified would be the route to success. He also set bait traps for the bark beetles around the outskirts of the town, which stopped many of them from reaching trees in the centre. The elm-tree 'buffer' of the South Downs was also extremely helpful. National statistics show that this strategy paid off and Brighton and Hove, together with neighbouring Eastbourne, now boast the finest collection of elms in Britain – designated as The National Elm Collection. Visiting tree enthusiasts can now enjoy a huge variety of English elms which, until 2019, included Britain's largest specimens, the famous Preston Park Twins. Tragically one of the twins was found to have Dutch elm disease and had to be felled. It is to be hoped the other 400 year old tree will survive. Brighton can also boast many other elm species, varieties and cultivars from every corner of the globe. In a cruel twist of fate the gales of 1987 and 1990 ripped many large and healthy trees out of the ground but Greenland and his team have tidied up and replanted and are dedicated to go on protecting their elms.

ALL IS NOT LOST

The common perception is that elms have totally disappeared from the British landscape but this is not true. Large trees may indeed have gone in most parts of the country but careful inspection of the hedgerows will reveal that it is still very much alive and well. English elm, which took the biggest hits from Dutch elm disease, is often so dominant that it has taken over in some places to create single-species hedges.

Different factors will shape the future of elms in the wake of the disease. Viruses may predate upon the fungus or even upon the beetles. Larger invertebrates may develop a taste for bark beetles. Global warming may affect the insect or fungus lifecycles. More elms may develop a resistance to the disease. Whatever the mode of survival or renewal, evolution will bring more large elms back into the landscape one day. Researchers are also trying to determine how some trees are able to resist disease; while new disease-resistant elm clones are currently in the process of trialling, mainly to see what their rates of survival will be, but also to find out what sort of shape they will develop in maturity. It's no secret that many people are hoping for a tree that will eventually adopt a form similar to the distinctive English elm profile; a tree that will really make that visual impact in open countryside. The big risk here is that when a clonal elm fails then so will most of its brethren. The renaissance of the elms is inevitable. Exactly when and how is difficult to predict.

A view down into the Olchon valley on the Welsh Borders shows a network of centuries old hedges, many of which have taken on the character of linear woods with the wealth of mature trees that now dominate them. While there is a practical stockproof barrier beneath them they now also provide valuable green corridors for the passage of wildlife.

Hedges and Hedgerow Trees

One of the most defining elements of the British lowland landscape is its remarkable network of hedges and hedgerow trees, complemented in the uplands by many miles of beautifully crafted dry-stone walls. These traditional features are an important part of our cultural heritage and lift the spirits and reaffirm the unspoilt pastoral glory of the countryside.

I never had noticed it until
'Twas gone, – the narrow copse
Where now the woodman lops
The last of the willows with his bill.

It was not more than a hedgerow overgrown.
One meadow's breadth away
I passed it day by day.

FROM 'FIRST KNOWN WHEN LOST'
BY EDWARD THOMAS (1878–1917)

The flowers of the exceptionally rare Plymouth pear in a hedgerow on the outskirts of Plymouth.

Hedges are some of the most significant features mapping the evolution of our landscape, steeped in history, part of our culture, wildlife havens of paramount importance. Some of the earliest-written records relating to hedges are to be found in the ninth- and tenth-century Anglo-Saxon charters, but there is no exact reference as to whether these were live or dead hedges – dead hedges being structures composed of cut stakes with interwoven branches. During the medieval period much of Britain's woodland began to be cleared to make more land available for farming, and strips of woodland edge, with plenty of well-established trees, were often left as hedges. Such hedges are most readily recognized today by their wealth of attendant ancient woodland-indicator plants, such as primrose and wood anemone, as well as the unusual appearance of trees more often associated with ancient woodland, such as wild service tree (*Sorbus torminalis*) or small-leaved lime (*Tilia cordata*). These trees have long since lost the ability to reproduce from seed because of climate changes and have had to rely upon regenerating naturally by suckering.

Secondary hedges have also evolved in places where neglect or lack of rigorous management has allowed the plants to colonize, typically along trackways and abandoned railway lines and around old fences and walls. Left to their own devices these may eventually develop into linear woodlands or, if managed, may become useful hedges.

There are many miles of deliberately planted hedges and most, over the last 250 years, have been rather uninspired hawthorn hedges, sometimes with blackthorn; both species create dense, thorny and effective barriers. These are typical Enclosure hedges. However, prior to the eighteenth century, hedge planting was principally undertaken with young woodland trees and set into new hedgerows with initial protection from stock by dead hedges or ditches. Although often similar in appearance to woodland relict hedges they don't tend to have such a rich herb layer beneath, being devoid of woodland-indicator plants.

Hedges were and indeed are usually planted for two main reasons: to mark the ownership of land, including public rights of way, and to control livestock. They also established boundaries that fulfilled a legal and social purpose. Respective rights and duties of tenure were clearly understood by all, since hedges were potential sources of timber, animal fodder, fuel and food it was important that communities had a clear picture of what was permissible and what was not. The Concise Oxford Dictionary defines a hedge as a 'closely planted row of bushes or low trees, especially forming [the] boundary of field, garden or road; similar boundary of turf, stone, etc.'. Broadly speaking, this is a fair description but since the trees are qualified as 'low' it fails to account for the taller individuals that are a common feature of many traditional hedges. Perhaps a better definition is the one that John Stokes of The Tree Council has come up with: 'a linear landscape feature consisting of managed woody shrub and tree species, forming the boundary of fields, gardens and roads'.

DIFFERENT TYPES OF HEDGE

Within the confines of Britain there is a remarkable diversity of hedge types, sometimes governed by the tree species they contain,

The famous double avenue of common limes at Clumber Park in Nottinghamshire.

A fine native black poplar in one of its hedgerow haunts in the Vale of Aylesbury. Hedgerows, preferably above watercourses, and riverbanks are the perfect habitat for this now quite rare native broadleaf tree.

sometimes by their physical characteristics or their historic purpose in the landscape and even affected by local traditions of planting and management. A typical British hedge will contain six or seven different species (maybe more on alkaline soils and less on acid soils) and the particular mix will vary from one part of the country to another. Certain species are notoriously poor colonizers so, as a rule of thumb, the presence of field maple, spindle, dogwood, holly or hazel frequently indicates a pre-1700 hedgerow. If they are all there with two or three other tree species, then a medieval origin is more than likely. The subtleties of regional mixtures of species is a massive study to contemplate but it is worth singling out some of the more clearly defined and unusual regional hedge types that often stand out in the landscape.

The 'hedges' in Cornwall, for example, are massive structures formed by a series of rounded granite boulders wedged together at the base, with layers of smaller stones above. Toppings of gorse, tamarisk, hawthorn and blackthorn are the most common. The

Herringbone or chevron pattern of stone construction in a Cornish hedgebank.

whole structure is knitted together with an earth core or infill, which makes them very stable, the width at the base being roughly double that at the top. Some of them have side walls constructed of smaller shales set in a herringbone or chevron pattern known locally as 'curzeyway' or 'Jack and Jill'. The hedge banks in Devon are similarly impressive and they often feature rows of large beeches, many of them wonderfully windblown, especially near the coast and on the moors. The beech hedges around Exmoor have clearly been shaped by severe weather over the years and many of them contain long, horizontal boughs, which are evidence of traditional hedge-laying techniques over several centuries. Hedges containing wild fruit trees, such as damsons, bullaces (wild plums) and sloes, were once an indispensable source of food in rural communities. Damsons are traditionally associated with the borders of Shropshire, Worcestershire and Herefordshire. They make excellent jams and wines and they used to be transported by the ton to the cotton mills of Lancashire for their blue-purple dye. Bullaces were once common in the hedges of east Hertfordshire and west Essex, although some of these are thought to be hybrids or seedlings derived from cultivated varieties. The sloe or blackthorn has a wide range, generally occurring as part of a mixed-species hedge. It is usually one of the first to show in late March (only wild plum preceding it by a couple of weeks), the pure white flowers appearing before the leaves. It forms dense, low-lying growth, which is good for nesting birds, and its wickedly long, sharp thorns make it the perfect choice for keeping livestock in (or out). If left unchecked it will throw up suckers and spread quickly. This suckering can occur along large sections of a hedgerow and may indicate a clonal community derived from one single plant.

The Plymouth pear (*Pyrus cordata*) is one of Britain's rarest wild fruit trees. Its range in Britain is confined to the West Country, though it is widespread in the rest of western Europe. The naturalist T. R. Archer Briggs first discovered the tree in Devon in 1865 and he observed that it was different from the other wild pears that he had seen, flowering later and producing fruit of a different shape. Whether or not this is a native species is debatable and it is difficult to explain why it might have been introduced, as the flowers smell disgusting and the fruits are virtually inedible. The species now enjoys special protection under Schedule 8 of the Wildlife and Countryside Act 1981.

A hedgerow tree that has attracted a good deal of interest in recent years is the native black poplar (*Populus nigra* subsp. *betulifolia*). This had been quietly growing in Britain, barely heeded by anyone until 1973, when Edgar Milne-Redhead decided to undertake a national survey after he retired as deputy keeper of the herbarium at Kew. Within a couple of years he had located about 1000 trees, which is a very small number indeed for such an important native broadleaf. The dendrologist John White spoke about the tree in a programme on Radio 4 in 1994 and then Peter Roe of the Daily Telegraph wrote an article about it and announced the launch of a nationwide black-poplar hunt. There are currently about 7000 specimens recorded, although there appears to be none in Scotland. There are three main reasons why this tree is so rare: the floodplains that they favour have declined because of agricultural drainage schemes; the timber appears to have little commercial value; a huge imbalance between male and female trees. In fact female trees account for less than 10% of the tree population, making natural regeneration from seed very difficult. Until recent times when concerted efforts have been made to plant female and male trees in close proximity the species has been left to its own devices for more than a century to regenerate naturally, usually from broken twigs that float down waterways, embed themselves

Damson trees in hedgerows occur as very localized features. These trees on the Herefordshire and Worcestershire border were once grown for the dye derived from the fruit and used by the Lancashire cotton industry. Now they are an early source of nectar for bees and provide delicious harvests for jam and wine-makers.

Edgar Richards of Beguildy hedge laying on the Powys and Shropshire border, using the traditional bill-hook.

in wet mud and strike roots. For this reason most black poplars tend to be found near rivers and streams and they do well in hedges that have ditches. The black poplar's profile is very distinctive, with its leaning bole and graceful, arcing boughs, and one of the best areas to pick it out in the landscape is in the Vale of Aylesbury.

Holly (*Ilex aquifolium*) has always been an excellent choice for the hedgerow. It is tolerant of a wide variety of soils and conditions and it forms a dense, stockproof barrier. It regenerates well after browsing by animals and has been used in the past in some areas as winter fodder. Holly features regularly in hedges around the Pennines, extending into Nottinghamshire, Warwickshire, Staffordshire, Cheshire and Herefordshire. Large trees often sit above the hedge line and these, according to local folklore, should be left intact in order to stop witches running along the tops. A more prosaic reason might have something to do with the fact that holly is a useful Christmas crop. A classic story from John Evelyn bears retelling. In 1664, Czar Peter the Great came to

England to learn about shipbuilding. He was staying at Sayes Court near the naval dockyard in Deptford in southeast London as Evelyn's tenant at the time and for a bit of fun he told his courtiers to push him through a holly hedge in a wheelbarrow. Evelyn was not amused, as he noted in *Sylva*:

> Is there under heaven a more glorious and refreshing object of the kind, than an impregnable hedge of about four hundred feet in length, nine feet high, and five in diameter, which I can shew in my now ruined gardens at Say's Court (thanks to the Czar of Moscovy) at any time of the year, glittering with its armed and varnished leaves? The taller standards at orderly distances, blushing with their natural coral: It mocks the rudest assaults of the weather, beasts, or hedge breakers.

Evelyn's beautifully detailed plan of his garden at Sayes Court (currently held at the British Library in London) dates from the early 1650s. It has a key describing every aspect of the house and gardens and the hedge in question is plainly marked as 'the hollye hedge' at the side of the mount or terras. (See also Orchards.)

If the glowing red berries of holly are a cheering sight in the winter landscape, then the effect of billowing banks of yellow laburnum flowers in early summer will lift the spirits too. Laburnum (*Laburnum anagyroides*) hedges are a strange though beautiful addition to hedgerow diversity. There are three notable examples: in Cumbria around Arlecdon, Haile, Ennerdale and Kinniside, on the Stiperstones of Shropshire, and in Cardiganshire in west Wales. Nobody knows why they were planted, especially as the seeds are poisonous, which makes them particularly unsuitable for fields containing animals. There is a slender theory that such hedges were vernacular understudies to the impressive laburnum hedges and walks to be found in some grander formal garden designs. The hard, chocolate-coloured heartwood is valued by cabinet-makers and it has been suggested that it was once used for the handles of miners' tools because it is similar to hickory. It is recorded that in the reign of Edward IV the wood was also used to make longbows, it being known as 'awlune'.

Yew (*Taxus baccata*) appears occasionally in wild hedges, most often as individual specimens, many of which are old boundary markers. Tree expert and writer Andrew Morton draws the conclusion from his observations in Shropshire that they served as guide trees to show people to the nearest church. Because yew is toxic to livestock and grows so slowly it would not have been an obvious choice for the hedge planter, so it's quite likely that big old yews were simply absorbed from the surrounding landscape. There are some especially impressive examples to be seen in the gardens of grand houses, most notably at Brampton Bryan on the Herefordshire–Shropshire border, at Bathurst Estate in Cirencester and at Powis Castle near Welshpool. In a tradition going back to the late seventeenth century, yew hedges were planted to create garden 'rooms', which offered privacy as well as protection for more delicate trees and plants. Hedge mazes were an innovation of the Elizabethans and were made purely for entertainment. If yew was a favourite with gardeners, then so was box. Both also made an excellent medium for the topiarists and such hedges, bearing all manner of birds, beasts and geometric fancies, are still widespread. (See also Yew.)

There is a strange knot of hedges, or deal rows as they are called, in the Brecklands of East Anglia, where Scots pines have been chosen for hedges or windbreaks for at least the last 200 years. One has to say that although the Scots Pine has a fine sculptural quality it makes a totally inadequate hedge. Maybe these deal rows should be classed as avenues rather than proper hedges. (See also Pine.)

Ultimately it is hawthorn that has been the most commonly employed hedging species, its thorny dense growth making the best of stock-proof barriers. It also provides perfect habitat for nesting birds, an abundance of fruit on which they feed in winter and makes a beautiful contribution to landscape character with a profusion of creamy white flowers in May – hence the alternative name for the tree. In some remote parts of Britain ancient gnarled and twisted trees may be discovered in fragmented remnant hedgerows. Particularly in exposed upland settings these relatively small trees may be several hundred years old.

The grand avenues normally associated with large country estates came into fashion in the late seventeenth century and they were planted to guide people through the parkland from the main house towards a feature such as a monument, summerhouse or lake. This was tree planting to impress. Most avenues were made up of a single species rather than mixed, the most popular trees for this purpose being oak, beech, elm, horse chestnut and – more often than not – the common lime. These regimented avenues fulfilled much the same function as the traditional hedge: a more symbolic demarcation but still a declaration of ownership. During the eighteenth century many of these formal plantings were swept away in favour of more open vistas with specimen trees and clumps. Research reveals that these so-called 'natural' effects were often achieved at the expense of vast areas of the real countryside, with farms destroyed, hedges uprooted and sometimes whole villages being evacuated to make way for a landscape that was deemed to be more pleasing to the eye. Many of the huge old pollard trees to be seen in parkland can be sighted into straight lines, making them almost certainly the remnant hedge trees of some long-lost hedgerows.

Although only about one third of a mile long, the Meikleour beech hedge stands 120 feet tall at its northern end, making it the tallest hedge in the world.

DATING A HEDGE

With the publication of *Hedges* by Pollard, Hooper and Moore in 1974 it looked as if the tricky business of dating hedges was going to get a little easier and more reliable. This book was one of the first fully rounded studies of British hedges and the authors present a scientific, historic and socially balanced overview of the subject. They concede that changes in agricultural practice had had an inevitable effect upon British hedges and proposed that the way forward would be through dialogue among the public, farmers, conservationists and the government. Forty-five years on this has started to happen.

One of the most memorable and frequently quoted features of the book was Dr Max Hooper's 'Rule' for dating hedges. He based his theory on the study of some 227 different hedges from around Britain, that could be cross-referenced to the Anglo-Saxon charters. Put simplistically, his conclusion was that the number of tree species (including wild rose) found in a 90-foot section of hedge corresponded to its age; so ten different trees, for example, would indicate that the hedge was approximately 1000 years old. How tidy and convenient this all seemed. Many landscape historians eagerly embraced the theory but it was not long before it came under attack. Hooper had admitted at the time that factors associated with soil types, climate, land use, later introduced species and the fact that woodland relict hedges could appear much older than they really were had led him to apply a caveat of plus or minus 200 years to his formula.

By the law of averages, and due to some convenient field studies, allied to maps and known hedgerow history, there were some hedges that fitted his formula. This suited some localized researchers admirably but more recent opinion has tended to acquiesce only with the broader views that planted species-rich hedges, if they can be dissociated from woodland relict hedges, are generally older because they have had longer to acquire a wider variety of tree species, either from deliberate planting or natural colonization, and not because they were necessarily initially planted with many species. To throw confusion into the equation, it must be remembered that older hedges could also lose species because they died out, were removed or crowded out by very successful suckering colonizers like the elm. Right up to the twentieth century it was popular practice in well-wooded areas outside the central belt of Enclosure to plant hedges with trees taken from the woodland, mainly because it was convenient and cost nothing. It is generally accepted that the more recent hedges are those containing fewer species – typically the Enclosure hedges of thorn from the last 250 years. Documents and maps are valuable for dating but only where a chronological sequence of specific landscapes can be studied. The earliest map of a hedge will only reveal its presence at that time, giving no indication of how long it had already existed. Studying field shapes on maps may be helpful: sinuous ones, following the shapes of old 'ridge and furrow', usually indicate medieval origins at least, while the ruler-straight thorn hedges cutting right across the old 'ridge and furrow' systems are invariably post-1700.

Ultimately, it is very difficult to ascribe an exact date to the majority of hedges. Certain features or species content are excellent hints and indications, and background documentation can be helpful, but hundreds of years of human activity, change of land and hedge use, as well as climatic influences, have caused hedges to evolve in many different ways all over Britain. The experts are still divided on many aspects and much research, both national and regional, continues. As Gerry Barnes and Tom Williamson note in their book *Hedgerow History*, published in 2006: 'The popularity of "hedge dating" is tied up with deeper issues, relating to the antiquity of the countryside, and to man's relationship with nature. For hedges, more than perhaps any other feature of the countryside, are both nature and history.'

A VERY BRIEF HISTORY OF LAND ENCLOSURE

By far the earliest evidence of field boundaries comes from remnants dating back to the Bronze Age, both in Cornwall, on the Land's End and Penwith peninsulas, and in Devon, where clear signs of the ancient reaves or walls stretch for many miles across Dartmoor. Whether or not these banks or walls were ever topped with any kind of live or dead hedge is unknown but it seems likely that they were. Remarkable aerial photographs of the reaves show a network of co-axial field systems, dotted here and there with tiny circles, which are all that remains of the dwellings once associated with these enclosures. Such patterns are very familiar today both in existing 'ancient countryside' and visible beneath some areas of eastern England, where 'planned countryside' overlies medieval systems, which in turn had ousted Roman field systems. Medieval open-field systems are most readily familiar today by evidence of the 'ridge and furrow' patterns, along with the sinuous reversed-'S' field shapes formed by the turning of oxen ploughs. Tight knots of fields for enclosing stock lay around the immediate fringes of villages, whilst the rest of the countryside was divided up into strips of land for growing crops, with individual ownership of the various strips, known as sellions, spread around several different locations. There might also be open commons where sheep could be grazed. By this time the economic importance of hedgerows was beginning to be a major consideration, and there are many records from the fourteenth century onwards concerning the costs of hedgerow planting, the value of timber and fuel wood and the harsh punishments and fines exacted upon those transgressors removing timber, firewood or even whole trees without permission. There are many fascinating and detailed references to these misdemeanours in contemporary documents and, at the time, these were matters taken extremely seriously. A few days in the stocks, a public flogging or a large fine was never to be taken lightly.

From the late Middle Ages more and more land became enclosed and after 1604, when the first Enclosure Act passed through parliament, the process gathered speed, reaching its zenith during the late eighteenth and early nineteenth century. Whereas land use had been subject to local agreements and a fairly even-handed distribution within the community, the Enclosure Acts reapportioned this fragmented allocation of land as larger individual self-contained parcels. It was a costly and time-consuming business to get an enclosure award petitioned and passed. Typically, it was the more privileged members of society who benefited, for many smaller owner-occupiers could not afford the costs and often sold out to their more prosperous neighbours. Landowners built hedges and fences to keep out hoi polloi, not only from their newly defined estates but often also from what had been common land. As a result, cottagers and labourers frequently lost out on their grazing rights.

It has been estimated that 200,000 miles of new hedges were planted between 1750 and 1850. Around 5000 separate Enclosure Acts enclosed over 7 million acres of open fields and commons. Seemingly endless straight hedges, almost exclusively of hawthorn, carved up the central belt of England, from Oxfordshire through to East Riding of Yorkshire, an area where almost 50 per cent of the open-field land became enclosed. Quick-sets (hawthorn saplings) were used and many nurseries grew rich on the sales of these hedging trees alone. Evidence of these large-scale grid patterns can be seen on the great wheat 'prairies' of the East Midlands, where maps and photographs will reveal how the Enclosure hedges forged their way relentlessly across old boundaries, field patterns and even roads. The whole countryside was totally restructured.

Meanwhile, much of southeast England and the West Country, including the Welsh borders, was relatively unaffected by parliamentary enclosure. Certainly enclosure had been ongoing for centuries as communities and individuals brokered deals that gave them more efficient use of the land available in relation to the needs of their farming specialities. A tradition in these regions had long been to take saplings from existing woodland for hedging. A mixed hedgerow was seen as a great provider – oak for construction, ash for fuel, elm for fodder, and fruit and nut trees for food.

After about 1870 Britain endured a succession of agricultural slumps, largely due to cheap imports of meat and grain. The lack of intensive management was beneficial to hedgerow wildlife resulting in a dramatic increase in the number of hedge trees from about 23 million in 1870 to 60 million by 1951. However, it wasn't just the turn in the tide of agriculture that helped hedges but also the increased use of fuels other than wood, principally coal, and the introduction of modern building materials, which reduced the country's reliance on oak. Hedges were often allowed to deteriorate to a point where they were gappy and full of dead wood, so they no longer provided stockproof barriers. Barbed wire, which arrived from America in the 1880s, was a relatively quick-fix solution, much cheaper and less labour-intensive than hedge maintenance.

From the 1950s there was a revolution in the whole style of farming. Mechanization stepped up several notches, particularly in arable farming. Ever larger agricultural machines were built to be more cost-efficient and labour-saving. To accommodate these, the fields also needed to be larger and have fewer boundaries. The obvious obstacles to this progress were the hedges, for not only were they physically inconvenient, but farmers also didn't want the shade they cast on the crops, or the vermin within them that ate their profits. Between 1945 and 1990 some 240,000 miles of them were grubbed out, an estimated 50 per cent of all of Britain's hedges. The new machinery was frighteningly efficient too. Chain saws and massive bulldozers could destroy a hedge in no time. The custom of stubble burning, banned in 1989, also took its toll. And when Dutch elm disease began to take out millions of hedgerow elms in the late 1960s there was little or no concerted effort to replace these with other species. The losses to the landscape and wildlife habitat were staggering.

However, since the 1980s the balance has been tipped back in favour of hedges, albeit slowly. Fortunately, a few conservationists, concerned politicians and responsible landowners have realized what was at stake. Grants and subsidies have been awarded for the planting of new hedges and the restoration of old and neglected ones. The lands of the grain barons will never look much different but elsewhere in Britain many hedges are now healthier than they have been for a long time. Substantial grant and stewardship schemes encourage farmers to look after their hedges for their wildlife conservation value and many of them are now willing to leave headlands around their crops to minimize the damage that was caused by deep ploughing in the past. In 1997 the Hedgerow Protection Act was introduced, which has stopped a lot of unnecessary destruction of hedges, since a good case must now be put forward to justify their removal. The future for hedges looks bright, and it is thought that there are now probably more being planted than removed, something unknown for perhaps the last 200 years.

THE FUTURE FOR HEDGES

With the renewed interest in hedges has come a revival of the skilled craft of hedge laying. It will never be a cheap option to lay a hedge, as it is infinitely more time-consuming than running a mechanical flail across an outgrown hedge, but the result is vastly superior. When a hedge has been allowed to grow too tall and all the bottom has become gappy and useless as a stockproof barrier then laying is the best solution, and the time to do this is in the winter months, between October and March, when there is little sap rising. There are many different regional styles – two of the most common being Midland and Welsh – but the general principal is the same. Some of the stouter trees are cut to hedge height, or stakes are driven into the ground at regular intervals along the hedge line. Oblique cuts are then made at the base of the other tree stems, which are carefully bent over so as not to break the live-tissue connection at the heel of the cut. These stems, called plashers, are then woven between the stakes, and pushed firmly down to make a dense, tight, live fence. To stop these springing up a braided ethering or binding of hazel is tightly woven across the top of the stakes. Initially stark, as the first surge of spring sap bursts the trees into leaf the healthy state of a rejuvenated hedge is clearly evident. An expert hand creates a true work of art and a hedge which will be good for several decades.

Hedge layers are growing in number. Agricultural colleges and conservation groups are running courses and some existing professionals are taking on apprentices. The sense of pride in a job well done is epitomized by the frequent and well-attended regional hedge-laying competitions leading, of course, to the annual national competition where all the top dogs fight it out to be the best. The use of tractor-mounted flails does cause much consternation to the public when it is misplaced or mistimed. This is the wrong method for outgrown hedges, where the result of trying to trim trees that are too big ends up looking like a battle-field. A broken-down hedge has poor habitat value and is open to disease. A regular light flailing of well-structured hedges, prefer-ably into an 'A' or flat-topped 'A' profile is best for wildlife and can be effective and look good, as long as it's not done too often, giving a chance for the new growth that bears autumn nuts and berries for the birds. Flailing after mid-March, when birds are nesting, is illegal.

One major drawback of hedge flailing is that many operators fail to lift their machinery off the hedges when they reach saplings or small trees, which is resulting in a death of young hedgerow trees. The knock-on effect of this will be that eventually, when all our older hedge trees have died and gone, there will be no replacements growing up in the hedges. This will bring radical changes to the appearance of the landscape, be seriously detrimen-tal to wildlife which relies on such trees, as well as reducing shade and protection for livestock.

In the mid 1990s the Council for the Protection of Rural England was keen to promote hedgerow conservation and, in particular, to encourage retention of the saplings that would develop into the next generation of hedgerow trees. Fifteen years ago The Tree Council picked up the same cause, but while their campaigns may have been high profile for a year or two the challenge now is to make sure that landowners and contractors continually maintain a responsible regime. Without hedges and their hedge trees Britain's landscapes would be visually bereft and the biodiversity and abundance of much of our wildlife could be seriously compromised.

Introduced Broadleaf Trees

It is estimated that there are currently around 2500 species of trees that will grow successfully in Britain (plus a vast number of hybrids and cultivars), though only 36 can be classified as truly native. Add to this figure around 20-30 whitebeam micro-species, various obscure elms and some of the tiny dwarf willows and it is still a very small proportion of species found in the UK. Some very familiar species, the sycamore, planes and chestnuts, for example, have been naturalized for so long it is hard to think of them as foreigners.

Hail, old patrician trees, so great and good!
Hail, ye plebeian underwood,
Where the poetic birds rejoice,
And for their quiet nests and plenteous food
Pay with their grateful voice.

Here nature does a house for me erect,
Nature, the wisest architect,
Who those fond artists does despise
That can the fair and living trees neglect,
Yet the dead timber prize.

ABRAHAM COWLEY (1618–1667)
IN PRAISE OF THE TORTWORTH CHESTNUT

The springtime leaves of the sunrise horse chestnut.

In many cases the first introduced broadleaves were specimen trees planted by the gentry on their estates. It is difficult to say, due to very few reliable accounts, what the position was up to the end of the sixteenth century, but the rate of introductions accelerated in the seventeenth century as the first plant hunters travelled overseas to collect new species.

THE FIRST COLLECTORS

Two of the first and most avid plant collectors in Britain were John Tradescant the Elder and his son John Tradescant the Younger. Tradescant the Elder, in his role as Charles I's gardener, made official visits to Russia, Spain and Algeria, always bringing back an abundance of plants, miscellaneous curiosities from the natural world and ethnic treasures. He planted his amazing collection in the garden of his Lambeth home, which became known as The Ark, and opened it to the public at 6d (2½p) a head. Tradescant the Younger inherited the post of royal gardener from his father and added many more species to the collection. He made three trips to Virginia between 1637 and 1654, returning with specimens of the tulip tree (*Liriodendron tulipifera*), the swamp cypress (*Taxodium distichum*) and the occidental (eastern) plane (*Platanus occidentalis*). The black locust (*Robinia pseudoacacia*) is also known to have grown in his garden, but may have come from France. Late in the seventeenth century Henry Compton, bishop of London, sent his missionary-cum-personal plant hunter John Bannister, to America. One of Bannister's most important finds was the magnolia (*Magnolia virginiana*) and the specimen he brought back was the first one ever seen in Europe. The bishop's magnificent gardens at Fulham Palace also boasted the first British imports of the black walnut (*Juglans nigra*), the scarlet oak (*Quercus coccinea*), the balsam fir (*Abies balsamea*) and the box elder (*Acer negundo*).

News of all these wonderful trees from overseas soon spread and people's appetite for ever more exotic species just grew and grew. And more and more plant hunters emerged to satisfy the huge demand. Among the most successful were two Quaker botanists, Peter Collinson in London and John Bartram in Philadelphia, who created an Anglo-American partnership that lasted for almost 35 years. Prince Frederick (son of George II) and his wife, Princess Augusta, were among the first to plant the exciting new species of trees reaching London – in the gardens of their estate at Kew. After Frederick died in 1751, Augusta continued with the pioneering work; her friend and fellow plant enthusiast the Earl of Bute once remarked that eventually her garden would 'contain all the plants known on earth'. This famed estate is now known as the Royal Botanic Gardens at Kew and some of the original trees that Augusta planted – known respectfully as Kew's 'old lions' – survive to this day.

THE GATHERING MOMENTUM

From the mid-eighteenth century plant hunters set out with more purpose than ever before, initially to the furthest reaches of Europe and the Caucasus and then to the east coast of America, bringing back all manner of exotic plants and trees. Among the new arrivals

The typical shape of a splendid horse chestnut in full flower.

were the Lombardy poplar (*Populus nigra* 'Italica') from Italy, the purple or, more commonly, copper beech (*Fagus sylvatica purpurea*) a sport of common beech originally discovered in a German forest, the manna ash (*Fraxinus ornus*) from southeastern Europe and the Turkey oak (*Quercus cerris*) from (big surprise) Turkey. A Suffolk gentleman called Mark Catesby accepted a commission of £20 a year to collect plants in America and in 1726 he returned with the Indian bean tree (*Catalpa bignonioides*) from Virginia.

By the beginning of the nineteenth century virtually no corner of the globe lay untapped. Plant hunters explored America's west coast and started to penetrate the more inaccessible nations, such as China and Japan. What had been a trickle over the previous 200 years now became a torrent and during the next hundred years or so the number of tree species known to the western world trebled. These were all propagated by the botanical gardens and nurseries and quickly released to an eager nation. It is interesting to note, however, that many of Britain's great landscapers, such as Charles Bridgeman, William Kent and Capability Brown, failed to seize the opportunity and stuck rigidly to the native species for their grand schemes.

The nineteenth century saw the arrival of the Japanese maples and flowering cherries, which became the decorative mainstay of parks and gardens all over the country. One of the most important botanists of the Victorian era, Joseph Dalton Hooker, brought back the impressive *Magnolia campbellii* and dozens of rhododendron species from the Himalayas, many of which would attain tree-size proportions in Britain. He was also responsible for introducing the first cider gum (*Eucalyptus gunnii*) from Tasmania; this is now the country's most commonly grown eucalyptus.

The nation's increasing enthusiasm stimulated a rapid expansion of nurseries in the nineteenth century and it was the wealthy nurserymen as well as patrons and collectors among the gentry who were able to fund more exploration. In 1899 the nursery of James Veitch and Sons dispatched Ernest Henry Wilson to Sichuan to bring back seeds of the handkerchief or dove tree (*Davidia involucrata*) – a species first discovered by the French missionary Abbé Jean-Pierre Armand David around 1870. The 13,000-mile journey to China was long and arduous and when Wilson finally found the single tree he had been looking for (the Irish plant hunter Augustine Henry had told him exactly where to find it), it had been reduced to a stump and a fine timber house stood beside it. One can only imagine his despair. He persisted and eventually found another specimen. What appear to be huge white flowers on the handkerchief tree are in fact the bracts around the flowers; these are visible at night, which helps to attract the moths that pollinate it. Wilson returned home in

Britain's biggest Turkey oak at Shute Barton in Devon.

triumph in 1902 only to discover that a French missionary, Père Farges, had apparently already brought seeds back to Paris some five years before.

DEVELOPING THE URBAN LANDSCAPE

Trees in the countryside have almost always served a practical purpose – as a source of building materials, fuel, food or game cover. This is not so important in an urban environment, where plants are often chosen for ornamental value alone. Not surprisingly, the first urban trees were planted for the benefit of the more affluent citizens; many of London's oldest trees are to be found in the West End, for example. With the Industrial Revolution came the mass migration of workers from rural areas to the towns, where most of them lived in cramped squalor. A few of the more philanthropic employers and forward-looking town planners decided to improve the environment by creating public parks and gardens for recreational purposes.

John Claudius Loudon was one such pioneer. He conceived a radical rethink of London's landscape and infrastructure in an article entitled 'Hints on Breathing Places for the Metropolis, and for Country Towns and Villages, on fixed Principles', which he published in his *Gardener's Magazine* in 1829. He proposed a green-belt system, the creation of more open spaces and an integrated scheme for public transport, food production and waste recycling. It didn't all happen in his lifetime but by the end of the century many of his ideas had been adopted. We now take the concept of the green belt for granted and Letchworth, Britain's first garden city, was completed in 1903.

Loudon was well aware of the influx of new trees and plants from abroad and he encouraged gardeners to use them in layouts

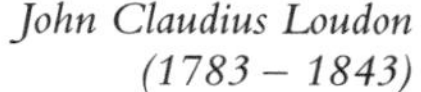

John Claudius Loudon (1783 – 1843)

based on abstract shapes. He called this style Gardenesque: 'When it is once properly understood that no residence in the modern style can have a claim to be considered as laid out in good taste, in which all the trees and shrubs employed are not either foreign ones, or improved varieties of indigenous ones, the grounds of every country seat, from the cottage to the mansion, will become an arboretum.'

Although the concept of an arboretum was well established, Loudon was the first person to use the term in English. Joseph Strutt, a local cotton-mill owner, commissioned him to design a public space so that the good people of Derby would have 'a Pleasure Ground or Recreation Ground'. Strutt envisaged a park that 'should comprise a valuable collection of trees and shrubs (from around the world), so arranged and described as to offer the means of instruction to visitors'. People were allowed to enjoy themselves but if they could learn something at the same time then so much the better. The Arboretum in Derby, generally considered to be Britain's first public park (and maybe the world's first arboretum), was officially opened on 16 September 1840.

Most of the large Victorian cemeteries that were built in London and other major cities from the mid-nineteenth century onwards doubled up as arboreta, an idea that Loudon also espoused. The capital saw the establishment of seven major cemeteries from 1832 onwards. Highgate, which opened in 1839, is perhaps the best known of these – Karl Marx, Michael Faraday and George Eliot are among the many celebrated names to be buried there. No longer just a gloomy repository for the famously dead, Highgate Cemetery has become a wonderful wilderness, rich in wildlife, though only a few elements of the original formal planting remain. Another one of the 'Magnificent Seven' is Abney Park in Hackney, created by Loddiges' Nursery. Conceived expressly as a burial place for Nonconformists, it boasted more than 2000 plant and tree species when it opened in 1840. This figure later rose to around 2500 species, which put it ahead of Kew at the time. Many of original plants came from the grounds of two grand houses that had previously occupied the site.

THE LONDON PLANE

There are some tree species that are such a familiar feature of our urban landscape that we invariably think of them as British natives. One of the most prominent examples is the London plane (variously *Platanus* x *hispanica*, *Platanus acerifolia* or *Platanus hybrida*) and its common name offers an obvious clue: it is estimated that they make up around 50 per cent of the capital's trees.

The London plane is a hybrid of two planes from opposite sides of the world: the oriental plane (*Platanus orientalis*), which is native

One of the huge old London planes in London's Berkeley Square.

from southeast Europe as far east as Iran, and the American sycamore, also known as the western plane or buttonwood (*Platanus occidentalis*), which belongs to the eastern states of America. The oriental plane first appears in British accounts as an entry in William Turner's *Names of Herbes* of 1548. It is believed that Sir Nicholas Bacon (the father of the writer and parliamentarian Sir Francis Bacon) planted the first one, in his garden in St Albans. The western plane probably arrived in the early part of the seventeenth century but records are fairly vague about when and where the original hybrid first occurred. Some authorities put it at around 1650 in southern Spain or France. The first botanical description of the tree in Britain comes from Oxford Botanic Gardens in 1670. It is known that both parent species were growing in the Tradescant garden in Lambeth in 1663 and since Tradescant the Younger had brought the American sycamore back from his travels to Virginia it seems very likely that the hybrid evolved there.

The London plane is one of the largest of the broadleaf species to do well in an urban environment. It thrives in most soils and will tolerate restricted root space and heavy pruning. More importantly, it can cope with quite heavy pollution, for it has tough, shiny leaves that are readily washed clean in the rain and its bark is able to renew itself by peeling off in small plates; this allows the breathing pores or lenticels to remain functional and clear of dirt. The characteristic flaking green-brown bark is one of the London plane's most distinctive features, along with the spiky round fruits that hang on throughout the winter. A mature tree can often grow to 131 feet tall with a spread of 82 feet, and the beautiful and valuable timber, known as 'lacewood', is much sought after in the timber trade. Some of the oldest specimens in London are more than 200 years old – the splendid trees in Berkeley Square, for example, have a known planting date of 1789.

THE BLACK LOCUST

British colonists first discovered the black locust (*Robinia pseudoacacia*) in the eastern states of America in 1607 and records place its introduction into Britain a few years later, possibly via France. It was named after Jean and Vespasien Robin, who had grown Europe's first specimens in around 1600 at the Jardin des Plantes in Paris, where they worked as gardeners and herbalists to Henry IV. Tradescant the Elder knew the Robin family and he may have been instrumental in bringing the tree to Britain.

The black locust is a member of the legume (pea) family. It has typical pea-like flowers, hanging in white sweet-scented racemes, and compound pinnate leaves. The tree contains several toxins and the young shoots bear sharp thorns, which give it a double defence against browsing animals. It can grow to about 82 feet and has a tendency to throw numerous suckers. This is brilliant for its survival in the wild but not so welcome in the garden. The cultivar

A fine engraving of the Common Robinia, locust-tree or false acacia from The Penny Magazine *of 1842.*

'Frisia', a bright yellow-green form first raised in 1935, is still a popular choice today.

There is an amusing story about the black locust tree concerning the English writer and social reformer William Cobbett, best known for his witty discourse on country life, *Rural Rides*. Cobbett was also a nurseryman and a successful entrepreneur. He decided to promote the tree as a 'new' discovery from America. This was the 'tree of trees', he told his wealthy clients, and its timber was far better than oak. His sales campaign was so successful that he couldn't keep up with demand. He imported tonnes of seed from America and when that ran out he resorted to buying plants from other sources. Luckily for him, most nurseries had adopted the common name of 'robinia' and so he was able to re-label the plants and get away with it. He imagined a curious writer in 200 years' time telling the world that 'The locust was hardly known in England until about the year 1823, when the nation was introduced to a knowledge of it by William Cobbett.' However, by 1838, Loudon had already debunked this particular notion. In Cobbett's defence it could be argued that he had travelled to America and had seen the tree growing in fine, straight stands. But no one as knowledgeable about plants as he could possibly have believed that it could outperform the good old English oak as a timber tree.

THE TULIP TREE

One of the largest and most important broadleaves of the eastern states of America is the tulip tree (*Liriodendron tulipifera*). The exact date of the first planting in Britain is uncertain but it was documented in Fulham Palace gardens in 1688 and it is believed that one coeval specimen still exists at Esher Place in Surrey; another example, which must be almost as old, grows in the centre of the old walled garden at The Hirsel, near Coldstream in Scotland. John Evelyn first mentioned the tree in *Sylva*, understanding it to have been introduced from Virginia by Tradescant the Younger, although he erroneously described this member of the magnolia family as a poplar. Indeed, the tree is often known as the yellow poplar or tulip poplar in the United States, the leaves having a similar tendency to tremble in the breeze like poplar. It probably gets its common name from its tulip-shape yellow flowers but some authorities insist it derives from the tulip-like profile of its leaves, which turn a superb buttery-gold colour in the autumn. It grows best in moist, fertile soils. It is not often chosen as a street tree in Britain, though it handles pollution well, and is more usually found in parks and large gardens. The timber of the tulip tree is easy to work and will steam and bend without losing strength. Early in the

A superb tulip tree in its autumn finery at the Royal Botanic Gardens, Kew.

twentieth century tulip wood was imported to make the frames of Morgan three-wheeler motor cars in Malvern (see also Ash). In America it was a traditional material for dugout canoes. It can grow to a prodigious size – over 115 feet high and 82 feet wide – and has been known to reach heights of 164 feet or more.

THE HORSE CHESTNUT

The British treescape would seem an infinitely duller place were it not for the common horse chestnut (*Aesculus hippocastanum*). No sticky buds bursting forth in springtime with creamy green down-clad leaves. No giant candelabras bearing the columns of exotic white flowers. No shiny mahogany conkers peeking from their spiky husks, the best of them always just out of reach to a stick-wielding schoolboy. This is a recreational tree in every sense. French naturalist Bom St Hilaire claimed that the tree was introduced into England from Tibet in 1550. The English botanist John Gerard described it in 1597 and reported seeing it in the Tradescant garden in 1633, where it had arrived from Vienna in 1616. The second half of the sixteenth century is now thought to be the most likely time, with Turkey cited as the most likely origin.

There are several explanations for the 'horse' part of this tree's common name. The most popular belief is that it refers to the horseshoe-shape scars that are left after the leaves have fallen but there is also some evidence that the nuts were fed to horses in Turkey. Horse chestnut timber is of poor quality and is not much favoured but the nuts have had their uses. Because they are rich in saponins, they produce an effective substitute for soap when crushed and mixed with water, and during the two world wars they were processed for their starch content to make acetone, which was needed to produce cordite for munitions.

It is not surprising that such a majestic tree became a favourite with the owners of grand country houses. Sir Christopher Wren designed one of the most spectacular horse chestnut avenues in the country, at Bushy Park near Hampton Court Palace. It was planted in 1699 and was 1 mile long. Queen Victoria first opened the park to the public in 1838 and this started the ritual of Chestnut Sunday, when people would gather to enjoy the best of the Maytime flowering. This lasted until the 1920s and it was started up again for Queen Elizabeth II's Silver Jubilee in 1977. The Sunday nearest to 11 May is the date chosen for Chestnut Sunday every year and the public can still enjoy this old tradition. Whether global warming will change this timetable remains to be seen.

There are several horse chestnut species and many hybrids and cultivars. The most familiar will be the red horse chestnut (*Aesculus* x *carnea*), which first appeared in Britain in the early nineteenth century. It is a cross between the common horse chestnut and the red buckeye (*Aesculus pavia*) from North America. This tree seldom achieves the stature of the common horse chestnut; it has no autumnal colour to speak of and is rarely more than 66 feet high. However, when the two species are planted together, the 'strawberry and vanilla' effect can be very attractive.

The Indian horse chestnut (*Aesculus indica*) is a bigger, broader tree than the common horse chestnut, with larger white flowers, which appear in midsummer. This tree was first introduced into Britain in 1851 from the northwest Himalayas, and in 1928 a fine cultivar from the tree, named 'Sydney Pearce', was raised at Kew. One of the more offbeat horse chestnuts is the gloriously named sunrise horse chestnut (*Aesculus* x *neglecta* 'Erythroblastos'), a cultivar that was developed at Behnsch in Germany in 1935. In early spring the rich salmon pink of the emerging foliage stands out as an exotic beacon in park and garden. In summer the leaves turn to a bright lime green.

THE SYCAMORE

The sycamore (*Acer pseudoplatanus*) is a controversial tree for it is so common and seeds itself so profligately that many people have come to despise it. Weeding it out is certainly a laborious job for the woodland conservationists who want to control its spread. But love it or loathe it, it is extremely well established in Britain. As The Revd C. A. Johns observed in his book The Forest Trees of Britain: 'No writer on the subject … looks on this tree in any other light than as a foreigner, but as a foreigner naturalized so completely that it will continue to sow its own seeds, and nurse its own offspring, as long as England exists.' On the plus side, the sycamore supports a wealth of wildlife: the leaf mould it produces is beneficial to earthworms and the flowers are a useful source of nectar for bees in early spring. The sycamore is first recorded in Henry Lyte's *A Niewe Herball* in 1578, although it is likely that it was introduced into Britain from southern Europe well before this date, probably by the Romans. Some people believe that migrating Celts brought the tree to Wales but William Linnard, in his Welsh Woods and Forests, published in 2002, claims it did not arrive until after the Norman Conquest. There is a carving of sycamore leaves on the shrine to St Frideswide in the cathedral of Christ Church, Oxford, dating from 1289. Maybe the stonemason responsible was an immigrant artisan who had seen the tree in Europe, perhaps in the grounds of a monastery, or maybe there were a few singular specimens growing in Britain at the time.

The name 'sycamore' points to a case of mistaken identity. Early reports show that it was confused with the *Ficus sycomorus*,

Loathed by many, the sycamore has many virtues and is an indelible part of the British treescape. In the 1970s, when Dutch elm disease eradicated so many elms from the landscape, especially from hedgerows, it was ash that filled the void. When ash dieback has done its worst sycamore will be the most likely species to colonise the new voids.

a fig species native to Palestine; the leaves of the two species are certainly similar. Its Latin name also signifies the earliest doubts of its identity – for it is the maple that is also a false or pseudo-plane, and again the leaves are not dissimilar to the planes. There are frequent early references to the great maple or false plane and in many parts of Scotland it is still known as the plane tree.

The upstart sycamore beats virtually every other tree in Britain for sheer guts and determination. It will grow vigorously in almost any type or quality of soil and will tolerate the most windswept of locations, from the salt-laden southern coasts to the exposed northern hills. Its distinctive winged seeds or keys germinate so well that a seedling can take root almost anywhere. I once observed a sycamore that had seeded itself inside an old willow pollard. In no time at all it grew so big that it eventually burst out of its host shell, ripping it apart from within.

Sycamores respond extremely well to pollarding or coppicing, producing an excellent white hardwood. The fact that it will not taint food has made it the traditional choice for wood-turned items such as bowls, platters, rolling pins and spoons. It is also currently popular with cabinet-makers and the best-quality, long, straight sycamore butts are sometimes more expensive than English oak.

Although the sycamore has spread to every corner of Britain, from Cornwall to Orkney, the most statuesque specimens are found across Scotland, frequently in the grounds of the great old ancestral houses. One tree at Newbattle Abbey near Edinburgh was thought to have been planted shortly after 1550, making it, until it was felled by a freak gust of wind in 2006, potentially the oldest sycamore in Britain. Which tree bears that mantle now? A somewhat dour reputation has attended a few of Scotland's sycamores, where they are known as dool or dule trees (an onomatopoeic word derived from the old Scots, meaning 'sorrowful' or 'mournful'), for these were hanging trees, a familiar feature in the landscape until the mid-eighteenth century. Set in a prominent position near a laird's residence, the sycamore with its dangling corpse was a stark warning to the locals.

One of the most celebrated landmark sycamores in Britain is to be found in the Dorset village of Tolpuddle. Here, in 1834, a group of six disgruntled farm labourers met beneath the tree and formed the very first trade union, an act that saw them arrested, sentenced and transported to Australia. There was public outrage and after petitions and protests they were pardoned by King William IV and the Tolpuddle Martyrs returned to their village three years later. The village green, with the tree upon it, was given to the National Trust in 1934; in 2004 they calculated that it was approximately 320 years old. This exceeds the lifespan of the average sycamore and its longevity is almost certainly the result of regular pollarding.

THE SWEET CHESTNUT

In historic terms the Martyr Tree is a mere infant in comparison to a sweet chestnut growing beside St Leonard's Church near the village of Tortworth in Gloucestershire. From a distance it looks as though one is approaching a small copse in a field but closer inspection will reveal that the woodland is actually all part of a single tree, for this gargantuan specimen has dropped boughs that have layered themselves all around the parent bole. Legend has it that the tree was planted some time during the reign of King Egbert (802–39). It is certainly documented as 'The Great Chestnut of Tortworth' in 1135. When Peter Collinson visited the tree in 1766 he recorded its girth at 52 feet and deemed it 'the largest tree in Britain', reckoning it to be 'not much less than a thousand years old'. Some loss of the central bole mass has since reduced its girth to 36 feet but even so this would probably make it Britain's oldest existing broadleaf tree (apart from some remnants of ancient small-leaved lime coppice stools said to be of greater antiquity).

The sweet chestnut or Spanish chestnut (*Castanea sativa*) is a native of southeast Europe and the Caucasus, and naturalized across the rest of Europe. The Romans valued it for its timber and edible nuts and it was thought for a long time that they were responsible for its introduction to Britain. However, Dr Rob Jarman of the University of Gloucestershire was not convinced by this theory, finding no evidence of its presence in the UK before AD650. He and his team spent several years undertaking DNA analysis of chestnuts in the UK and making comparisons with

An engraving from The Gentleman's Magazine *of July 1769 of the Tortworth Chestnut in Gloucestershire. It is thought to be around 1200 years old and thrives to this day within its own enclosure behind Tortworth village church.*

European data, publishing their findings in 2019. They concluded that British sweet chestnuts traced their genetic lineage back to trees from western Europe, rather than the southeast quarter that was the hub of the Roman empire.

Several features make the tree readily recognizable although Nicholas Culpeper informs us that 'it were as needless to describe a tree so commonly known as to tell a man he had gotten a mouth'. The deep green leaves are long, pointed and elliptical in shape, with regular pointed teeth along the margins. The bark in mature trees develops into a distinctive spiral or twisting pattern around the bole. The nuts, known to the ancient Greeks as 'Zeus's acorns', have spiny green husks and they used to be ground to make flour for baking into bread. The traditional method of preparing them for eating whole is to roast them on an open fire or brazier. Evelyn called the chestnut 'a lusty and masculine food for rustics at all times, and of better nourishment for husbandmen than cole [cabbage] and rusty bacon, yea, and beans to boot'. Culpeper acknowledged their nutritional value and he also recommended them as 'a good remedy to stop the terms in women' (that is to alleviate heavy menstrual bleeding) and 'an admirable remedy for the cough and spitting of blood'. Powdered nuts were once a country remedy for haemorrhoids. This belief almost certainly harks back to the Doctrine of Signatures, a medieval system of classification whereby the beneficial or curative properties of plants were signified by their physical characteristics. Britain's climate tends to produce nuts of inferior size and quality compared to the tasty marrons that come from a specially cultivated variety grown in Italy, France and Spain.

Sweet chestnuts can still be found in many of the woods in the southeast of England, especially in Sussex, Kent and south Essex. Sweet chestnut grows fast, a typical rotation for coppice regimes being 14 years, according to forester and writer H. L. Edlin.

Regular coppicing increases a tree's lifespan and there is evidence of this in the numerous impressive old coppice stools that remain in these woodlands. The huge network of woodlands around Canterbury, known as The Blean, is one of the very best places to see large areas of ancient sweet chestnut coppice. These are still harvested from time to time as part of the ongoing management of a very vibrant woodland system. The timber of the sweet chestnut is never as sound as oak, as it has a tendency to 'shake' or crack, usually along the lines of the annual rings, which makes it less desirable as a building material. However, it will cleave extremely well and is suitable for fences and posts. In Kent it was traditionally used for hop poles.

As an ornamental, the sweet chestnut has always been a popular choice for large gardens and parkland. The triple chestnut avenue at Croft Castle in Herefordshire is the only one of its kind in Britain. Established around the time of the defeat of the Spanish Armada in 1588, legend has it that British sailors salvaged chestnuts from Spanish galleons. The rows of neatly planted chestnuts were supposed to represent the linear squadrons of Spanish warships, whilst the ancient oak pollards scattered somewhat haphazardly all about them were the British vessels. Although it seems a shame to let the truth get in the way of a good story, there may be an alternative explanation. A group of tree wardens from the West Midlands visited Croft Castle a few years ago and someone came up with an interesting hypothesis: maybe the sweet chestnuts in the triple avenue were originally planted as an orchard. This is certainly quite feasible – the sixteenth- century poet and farmer Thomas Tusser included chestnuts on his list of fruit trees suitable for cultivation – and the grid pattern of the triple avenue does look similar to the layout of a typical orchard. There is so little accurate documentation of early tree planting that we will probably never know the truth in this case but the avenue certainly satisfies two of the principal reasons for planting trees – they were useful and were also handsome additions to the landscape.

THE FUTURE

There are many people in Britain who look askance at planting anything but native species outside the confines of a park or garden. But we live in a changing world and we can not put the clock back; anyway, our stock of so-called native species has already been corrupted to a certain extent by hybridization. Is the English oak still English or has it become a European like the rest of us? If international trade has brought about widespread changes, then these will seem insignificant when placed in the context of global warming and the effect that the melting of the ice-caps will have on the Gulf Stream, that helps to keep our climate mild. The critical timescale may be as short as 50 years. As well as a general raising of mean temperatures, it is predicted that weather conditions in Britain will become more extreme and prolonged periods of drought and violent storms have already begun to affect the nation's trees. We know, for example, that the shallow-rooting beech will suffer more in warmer conditions than some of the other species. As some native species give way, a combination of a new order of natural succession and intervention planting will create whole new treescapes.

My view is that there is no point in mourning our projected losses, for the course is already set. We should plan for the future in a spirit of constructive optimism rather than despair, and prepare for the inevitable by making some judicious planting decisions now. This means considering the many non-native species at our disposal, both for amenity purposes and for commercial forestry.

And with a base of more than 2500 suitable species from which to choose, the opportunities for adventurous tree planting in Britain have never been greater.

Of all the London planes in London the Wood Street tree, just off Cheapside, is arguably the most famous. Quite how it has survived for over 200 years in the midst of the cityscape is something of a miracle. This is a rare Victorian lantern slide of the tree from around 1890.

The Scots Pine

The Scots pine (*Pinus sylvestris*) has a vast range across Europe and beyond into northern Asia. Also commonly known as the Scotch fir until the end of the nineteenth century, it has spent the last 9500 years heading northwards in Britain. A subspecies called *scotica* grows in the Scottish Highlands and these pines are the only genuine natives.

This is the tallest tree within my woods,
Lean, rugged-stemmed, and of all branches bare
Full thirty feet, with green plumes in the air
And roots among the bracken. All his moods
Are rough but kingly; whether, grand, he broods
Above his full-leaved comrades in the glare
Of summer, or in winter, still more fair,
Nods princely time to the wind's interludes.

FROM 'THE SCOTCH FIR' BY WILL H. OGILVIE (1869–1963)

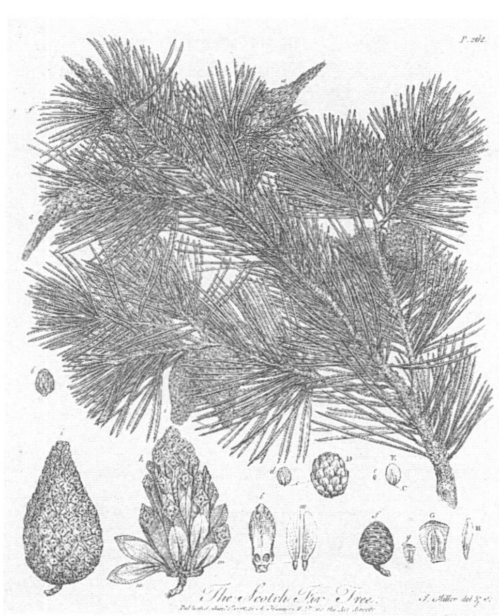

Typical of the fine botanical drawings by John Miller in the 1st Hunter Edition of Evelyn's Silva *is his depiction of the Scotch fir tree (the early name for Scots pine).*

Virtually all the Scots pines occurring further south than the Scottish Highlands have grown from non-native sourced seed or plants, usually brought in from continental Europe. There are several subtle differences: the trees of the subspecies tend to have a more rounded profile in maturity, with smaller cones, reddish bark and shorter needles of a decidedly blue or glaucous colour. They are tenacious and will grow on the poorest and rockiest of soils. Their somewhat rugged, shaggy form – old veterans are called 'granny pines' in Scotland – means that these trees have invariably been overlooked, both as an ornamental and as a commercial crop. Certain other evergreen conifers, such as the cedars and redwoods, will always look more impressive in a formal landscape while the Douglas fir, larch and Sitka spruce have all eroded its supremacy as a timber tree.

THE CALEDONIAN FOREST

One of the most hotly debated issues surrounding Scots pine and its role in the landscape history of Scotland is the conundrum of the Caledonian Forest, sometimes called the Great Wood of Caledon. It is widely believed that a vast tract of pine forest once stretched almost unbroken from coast to coast across the Scottish Highlands, perhaps covering more than 3 million acres. If this supposition is correct, then what remains accounts for barely 1 per cent of the original forest, amounting now to about 30,000 acres, most of which is centred upon half a dozen large sites and about 30 smaller pockets.

The best places to see what appear to be naturally occurring pinewoods, containing multi-generational stands of trees, from seedlings to veterans of perhaps 400 years, are: on Deeside at Glen Tanar, near Aboyne, Ballochbuie Forest, near Braemar, Glen Affric to the southwest of Inverness, on Speyside in the great forests of Rothiemurchus and Abernethy and the Black Wood of Rannoch in Perthshire. All these places seem to be ancient and natural but most of them have been extensively felled and replanted. The tradition of using fences to keep out the deer encourages the trees to regenerate naturally as does the burning off of certain areas, whether this is done deliberately as part of forest management or occurs accidentally through lightning strikes. Fire clears the ground of heather and peat, allowing the next generation of pine seedlings to gain a toehold in the gravelly soil beneath. Mature trees have a thick bark that can withstand fire well but when they do succumb they burn fiercely because they contain flammable resinous oils.

So, if there really was a great seamless pine forest, where is the rest of it now? It must be remembered that a considerable proportion of the Highlands is taken up by mountains, many of which are well over 2000 feet, and few pines will grow above that altitude. Ancient pines, sunk deep in the Scottish bogs, go back many thousands of years, suggesting that the same processes that ousted the Scots pine from England were also at work in the north. The increasingly wet climate made the habitat less suitable for the pines and in the south this was exacerbated by the overriding success of the colonizing broadleaf trees, beneath which pine seldom thrives. As blanket bogs dominated parts of the Scottish uplands the pines' struggle for survival intensified and the limitations of a suitable

Craggy old Scots pines in the Black Wood of Rannoch in Perthshire.

habitat narrowed down the likely locations for woods. If environmental changes have affected the pinewoods, then without doubt human intervention has also made a considerable difference too.

Several travellers to the Highlands in the eighteenth and nineteenth centuries have left detailed accounts of the great devastation of the forests there. Pines do not regenerate in a coppice-like manner as do many of the broadleaf trees, so the piles of brash (accumulation of smaller cut branches of no commercial use) stretching to the horizon must have looked really bleak. The Welsh naturalist Thomas Pennant toured Scotland in 1769 and this is what he saw as he approached Glen Coe from the south: 'A few weather-beaten pines and birch appear scattered up and down, and in all the bogs great numbers of roots, that evince the forest that covered the country within this half century. The pine forests are become very rare.' His published works were very popular and his accounts and opinions (right or wrong) were taken into common belief.

The English naturalist Prideaux John Selby visited the Highlands in 1824 and he published an account of his travels in 1842. He was shocked by what he saw in Glenmore, a valley situated between Deeside and Speyside:

> the tract previously occupied by this once magnificent forest exhibited a scene of savage wildness and desolation. Scattered trees, some of which were in a scathed and dying state, of huge dimensions, picturesque in appearance from their knotty trunks, tortuous branches, and wide-spreading heads were seen in different directions, at unequal and frequently at considerable distances from each other; the solitary and mournful-looking relicts of the departed glories of this once well-clad woodland scene, and which had only escaped the

> axe from their previous decay, or the comparative worthlessness of their knotty trunks, while the surface of the ground in almost every direction was littered and bristling with the decaying tops and loppings of the felled trees.

He noted that mosses were colonizing the area because the tree felling had interfered with the natural water courses and the ground was retaining more surface moisture; he feared that the land would eventually become a peat bog. However, just before he published his account the Scottish writer Sir Thomas Dick Lauder informed him that he had seen thousands of pine seedlings in the forest of Glenmore. The fact that the area was replenishing itself indicates that something must have happened to stop the natural process. The most likely explanation is that the landowners had removed the brash by burning it off. They may then have put fences round the area to keep out sheep and deer.

A red deer stag scenting a tree as a territorial marker.

Similar procedures of harvesting, planting and natural regeneration were going on in most of these pinewoods, and old records and maps show that over successive cycles the wooded areas may have moved slightly, yet seem to remain relatively constant in size. The pressures on land use were not restricted to forestry and in the wake of the Highland Clearances in the late eighteenth and early nineteenth century many absentee landlords introduced huge flocks of sheep and encouraged the red-deer populations for the hunt, all of which had a limiting effect on natural regeneration. Deer continue to be a problem for foresters today, as do the ravages of capercaillie. In essence then, if there ever were a vast pine forest across Scotland it wasn't the relatively recent ravages of human intervention that sealed its fate but the environmental changes that occurred thousands rather than hundreds of years ago. Woodland management in Scotland was always as skilfully conducted as anywhere else in Britain, so much so, in fact, that many nineteenth-century English and Welsh landowners went out of their way to hire a Scot to manage the forests on their estates. Timber was a valuable resource, so fencing off browsing livestock was a fundamental practice, and clear-felling without a view to replanting or nurturing regeneration would have been tantamount to commercial suicide. If observers of the Scottish landscape weren't bemoaning the lack of trees, then they were often equally depressed by the ranks of healthy even-aged forestry trees. As Selby pointed out, the only big trees that avoided the foresters' axes were the misshapen or decaying specimens; fortunately, some of those odd and uncommercial trees are now the veterans of today. Recent surveys on Deeside have revealed a handful of old-timers in excess of 400 years old.

One of the most beautiful places to see Scots pines at their untamed best is in Glen Affric, southwest of Inverness. In the spring the emerging emerald-green foliage of the birches set against the dark billowing heads of huge weather-beaten pines, as they hang from the precipitous crags, might only be rivalled by the autumn vista, when the birches have struck their golden glory – a visual overload for those eighteenth-century admirers of the picturesque. Further up the glen some fine old individual pines stand alone, the grace of their fully matured and rounded forms shown to best advantage in the first or last sunlight of the day, when the reddish hues of their great boles glow with a warmth that seems to come from within.

THE SCOTS PINE SOUTH OF THE BORDER

In England and Wales the Scots pine is common enough, yet it can not strictly be considered a native species. Most of the trees are forestry plantations, established at a time when there was no particular desire to adhere to the use of native seed stock. The trees have done well but the very best pine timber is still reckoned to come from the slower growing native stock of Scotland. The locations chosen as the most likely places for success were either inhospitable uplands where it wasn't too damp or, more often, areas of low-lying sandy soil.

Some of the largest plantations on these latter soils are in the Breckland of East Anglia, around Thetford, and these are useful commercially and as an amenity for the local populace. A more interesting pine feature of the landscape hereabouts is the numerous pine rows or hedges alongside fields and roadsides. They were mainly planted from the end of the eighteenth century and throughout most of the nineteenth century by local landowners; they were the best trees for the light sandy soil and they also provided shelter belts for crops and game. Known locally as deal

The pine or 'deal' rows of the Brecklands in East Anglia – strange hedgerow pines that never quite performed as intended.

rows ('deal' being the old East Anglian name for any conifer), some rows of trees grow perfectly upright, whilst others would appear to have once been laid into hedges many years ago. It is said that a lack of labour during the Great War left these hedges unmanaged and the trees quickly grew out of the hedge; those that survive have often been windblown into fantastic shapes, forming a unique regional landscape feature.

The Scots pine has naturalized itself across much of Surrey and Sussex, and certainly gives the appearance of a native species, yet when it grows on land grazed by livestock or where deer are prevalent it will suffer. Charles Darwin, writing in *The Origin of Species*, discovered a perfect example of the pine's struggle to survive on a Surrey heath near Farnham: 'On looking closely between the stems of the heath, I found a multitude of seedlings and little trees which had been perpetually browsed down by cattle. In one square yard I counted thirty-two little trees; and one of them, with twenty-six rings of growth, had during many years tried to raise its head above the stems of the heath, and had failed.'

Great ranks of commercial Scots pine have made way for other more versatile or productive conifers in southern Britain, and among the pine clan it is more often the Corsican pine (*Pinus nigra* subsp. *laricio*) that has displaced it. Corsican pines grow quicker, produce equally good timber and, ultimately, will be a superior choice for commercial forestry as global warming changes Britain's climate.

Pine clumps and pine-ways are to be found all over England and Wales, with particular concentrations along the Welsh borders. Documentary evidence explaining why these pines were first planted is difficult to trace, probably because it is too long ago, but long-standing tradition and regional hearsay suggests that they were meant to act as guide trees for sheep and cattle drovers. It makes sense. Dark, evergreen trees, which would grow in almost any soil type, even on the tops of prominent and rocky hills, would make splendid waymarkers at any time of year. It is also said that clumps of pines around farmsteads or fields indicated that a drover might find hospitality there and temporary pasture and enclosure for his stock. Dorothy Halliday, a contributor to *Flora Britannica*, notes that several Oxfordshire estates have pine avenues or pines planted around the farms, popularly thought to have signified safe houses for Jacobite supporters during the eighteenth century. It is an interesting theory but some corroboratory records are still required.

The visionary polymath Alfred Watkins mapped out the paths of ancient ley lines in the borders after he had a moment of enlightenment on 30 June 1921 when he was roaming the countryside near Leominster in his home county of Herefordshire.

Suddenly, a correlation of the numerous landscape features he saw about him and the patterns he could see upon his maps coalesced into a vast network of straight lines. Hill forts, churches, standing stones and many pine clumps rather conveniently sat in the paths of his ancient, unseen, straight tracks. Watkins believed these were the visual reference points along communication routes of ancient British peoples, accepted by custom down many centuries and later incorporated into the Christian pattern of faith by the early church builders, who sited so many of their places of worship upon the leys. How sensible to build your church in the direct path of travellers and prospective converts. Churches were safe havens in a wild and sometimes dangerous land. The existing pines today are seldom much in excess of 100 years old but the tree naturally regenerates with ease from seed so these examples may well be direct descendants from pre-history. It is also quite possible that pines have been regularly replanted, for it's remarkable how ingrained local superstitions and traditions can become.

THE USEFUL SCOTS PINE

The Scots pine has long been held in high esteem for building construction and for shipbuilding. The hard, strong, resinous wood is useful for floorboards, joists and rafters as well as pit-props, telegraph poles and fence posts. Selby mentions pine's use for building entire ships, and yet only for merchant vessels for, as he points out, the oak is superior for vessels of war as it is stronger and more durable than pine, which has 'the disadvantage of splintering to a much greater extent when struck by cannon shot, which is always attended with greater danger and destruction of life, than when the missile merely opens itself a passage through the wood'. More often pine was selected just for the masts and spars. There is a hill in Glen Affric called Beinn nan Sparra, which means 'Hill of Spars'.

A great deal of Scots pine has traditionally been imported into Britain from the Baltic. In the nineteenth century the timber was known as Dantzig pine or Riga pine, but more latterly as yellow deal, red deal, redwood or northern redwood. The best timber has always come from the colder climes, where the tree grows more slowly. In the eighteenth century pine logs from Rothiemurchus on Speyside were converted into water pipes, bored out in a similar fashion to elm. William H. Ablett, in his book *English Trees and Tree- planting*, published in 1880, describes how timber was extracted and transported from Rothiemurchus – the sort of sight we can only relate to the great logging activities in Canada today: 'In times of drought the workmen collect the trees in what they term a dell or den, and build up a temporary dam, and await the coming of a flood. When this takes place, and the temporary dam is full of water, the dyke is broken, and the trees are swept down with the course of the water to the Spey.' Often these logs would then be floated all the way down the Spey to the coastal shipyards and mills in Aberdeen.

The inner bark of pine was once woven into ropes, much as the bast (inner bark) of lime was used. The resinous products of the tree were either collected as turpentine or heated up to make tar. Ablett reported that the virtue of pines as a benefit to persons 'suffering from diseases of the lungs, by residing in the vicinity of pine-forests, has long been known'. Indeed, pine does provide a marvellous remedy for respiratory ailments, as it clears congestion; it relieves asthma, bronchitis and flu symptoms, as well as coughs and sore throats. It can be used as a remedy for arthritis and gout and may also relieve headaches and toothache. Pine may be used as an invigorating tonic or as a calming and refreshing

A clump of pines on a hill top above Swaledale in Yorkshire.

relief from exhaustion and stress. It is also to be recommended for the circulation and digestion. Used externally, pine resin and oil is excellent for the relief of muscular pain or aching joints. It also makes a good deodorant and insect repellent. All round, pine is an exceedingly useful tree.

THE FUTURE FOREST

Ultimately, it doesn't really matter what the reasons were for the disappearance of the Caledonian Forest. If it were climatic and many millennia in the past, then it couldn't be helped. If it were over the last few hundred years, then we seem to know why and can hopefully redress the balance a little. The most optimistic prospect for the future for the great pinewoods is to harness policies of responsible forestry with informed habitat management. Landowners in Scotland still divide their profits between deer and trees but most now realize that managing the numbers and range of the deer will benefit their pinewoods. It's a matter of fencing off pockets of land to allow natural regeneration, but fencing different areas at different times so that not all the trees are of the same generation. Exclusion of deer not only helps the pines but also benefits the other constituents of the habitat – plants, mosses, lichens, invertebrates and birds.

Trees for Life, based at Findhorn Bay in northeast Scotland, is dedicated to resurrecting the great pine forests and since 1989 this splendid organization has planted over half a million native trees; that's not just pines but aspen, juniper and rowan as well. They are also closely monitoring and encouraging natural regeneration of colonies of the tiny, scarce dwarf birch (*Betula nana*). The aim is to recreate 600 square miles of Caledonian Forest. Whether or not we shall ever again see the wild boar, brown bears, lynx, moose or wolves roaming these forests is an entertaining thought (maybe not so much for the sheep farmers) and it would reintroduce natural predators to control excessive deer populations. There have already been trial reintroductions of beavers, with the current numbers around 450, and while these animals have had a very positive effect on the landscape and biodiversity they haven't been so warmly welcomed by farmers, resulting in controversial culling orders in 2019 for a fifth of the population. It seems bizarre that an animal is re-introduced after more than a 400 year absence and then promptly culled before it has reached sustainable status. So who can say what the future may bring.

The weird contorted form of a windblown Scots pine on the beach at Yellowcraig in East Lothian.

Introduced Conifers

There are those who would decry the very mention of introduced conifers, summoning as they do the spectre of oppressive ranks of Sitka or Norway spruce. These factory forests may present an unbroken swathe of monocultural monotony, blanketing mountain and moor, but they can provide valuable habitats for certain wildlife and are a vital part of the nation's economy.

Serene the stately redwoods stand,
Untouched by time, nor slaves to fear;
High Elders of the golden land,
Tall guardians of its secrets dear.
No twist or knot, no strain or stress,
Disturbs their perfect symmetry,
They rise unmarred from base to crest,
In forms of pure geometry.

FROM 'SEQUOIA SEMPERVIRENS' BY DAVID HOFMAN,
TAKEN FROM 1947 ANTHOLOGY SPIRIT OF THE TREES

Larch 'roses' are the beautiful immature forms of the cones that bear the seeds.

The Forestry Commission owns most of the plantations of introduced conifers in Britain and it has recently begun to change its policies. It now embraces the principle of species diversity and has started to introduce broadleaf trees, which will enhance and diversify habitats for wildlife and please the public at the same time. Historically, the reason why the Sitka spruce became so ubiquitous is because it grows exceptionally well where others fail. Introducing conifers into the country has largely been driven by the search for profitable softwood species that would satisfy the ever increasing demand for commercial timber, and pulp for paper manufacture.

A handful of conifers came over from the Old World from the seventeenth century onwards but the most dramatic finds began to cross the Atlantic in the early part of the nineteenth century, particularly after the west coast of America was opened up. Plant hunters such as David Douglas are surely owed the greatest debt of gratitude for the fantastic variety of trees they obtained for the nation, almost all of which turned out to thrive particularly well in the British climate.

THE NORWAY SPRUCE

The Norway spruce (*Picea abies*), also known as the spruce fir, has long been one of Britain's most important softwood timber trees and this species has shaped great tracts of the landscape. The early sixteenth century would seem to be its most likely date of introduction, for it is first mentioned by Thomas Turner in his *Names of Herbes* in 1548. Its native range covers most of northern Europe, Scandinavia and western Russia. The name 'spruce' derives from the old state of Pruse or Prussia, which would have been one of its strongholds.

Fossil records show that prior to the great ice ages the Norway spruce would have grown in Britain but it was one of those species that never recolonized naturally after the ice retreated. In places like the Alps or the Pyrenees it will tolerate altitudes of at least 6600 feet, growing to a height of more than 197 feet in the more favourable locations. In Britain it achieves a more modest height of around 130 feet, although most of the trees will be harvested well before that.

In 2008 Leif Kullman, Professor of Physical Geography at Umea University, discovered a truly remarkable Norway spruce on the Fulufjallet Mountain in the Dalarna Province of Sweden. Superficially the extant main trunk was only a few hundred years old, but carbon dating of samples from the root system established that this tree was over 9,550 years old, its great antiquity achieved by a continuous cycle of renewal by layering and vegetative cloning from the ancient root system. 'Old Tjikko' (Kullman named the tree after his dog), can now claim to be the oldest Norway spruce and the third oldest tree in the world, albeit clonally. In addition some twenty other spruces in the same area were discovered to be around 8,000 years old. The oldest tree in the world is generally considered to be a quaking aspen (*Populus tremuloides*) clone in Fishlake National Forest, Utah. It covers an astounding 107 acres, has around 47,000 stems (constantly dying and renewing) and is the heaviest-known organism estimated to

One of the largest Sitka spruces in Britain is this massive specimen in Drumtochty Glen in Aberdeenshire. It is 165 feet high and has a girth of 22 feet 3 inches.

weigh some 6,000 tonnes. Its age is calculated to be 80,000 – 1,000,000 years old!

It is strange to think that the first Norway spruces were planted for their ornamental value. A few nineteenth-century botanists saw the potential of the species and recommended it for spars and masts for ships, scaffolding poles and ladders, floorboards and all manner of interior joinery, its advantage being that its stems grow up long and straight when the trees are planted close together. It also found favour as a material for furniture and musical instruments. Other parts of the tree have proved useful as well: the inner bark for baskets, the roots for cordage, the resin for a substance known as Burgundy pitch, the tender young shoots as an ingredient for spruce beer, which, when added to molasses and yeast, produces a fermented alcoholic beverage that sailors used to drink to ward off scurvy, and the fronds (often mixed with juniper) for an aromatic floor covering.

Industry continued to find new uses for the Norway spruce and by the beginning of the nineteenth century it was being turned into telegraph poles, pit-props and wood pulp to make paper. Turpentine was also refined from the resinous pitch. When the Forestry Commission was formed in 1919 it began to plant the tree extensively, especially in Scotland. It seems that Britain can never grow enough of this and the other softwood trees required to satisfy the country's needs and we still import huge quantities from Scandinavia and eastern Europe. This is difficult to grasp when you can stand in the middle of somewhere like Kielder Forest in Northumberland and see nothing but rows and rows of Norway and Sitka spruce stretching into infinity.

In spite of its dominance as a commercial crop, the Norway spruce maintains links with its ornamental past – horticulturalists now have more than 350 listed cultivars from which to choose. In a domestic setting it is perhaps best known as the Christmas tree. The tradition of bringing evergreen trees or boughs into the home in winter is pagan in origin and it is thought that this developed into a Christian custom in Germany some time in the sixteenth century. Prince Albert (a German) was responsible for popularizing the idea in Britain when an illustration was published in 1848 showing the royal family celebrating Christmas around a decorated tree. After that, of course, every household in the country wanted one. In 1947 the people of Oslo sent a huge Norway spruce to the people of London as a token of their gratitude for Britain's support during the Second World War. The tradition has continued and a Norwegian Christmas tree is erected in the middle of Trafalgar Square every year.

THE LARCH

There is no definite record of the European larch (*Larix decidua*) first arriving in Britain. King Charles I's apothecary John Parkinson mentioned it in 1629 but only as 'rare, and nursed up with a few, and those only lovers of variety'. John Evelyn noted its presence in Essex in his 1664 *Sylva*, regarding it as a 'goodly tree', and identified its potential as a productive source of timber. Nobody appeared to take much notice at the time and large-scale planting in Scotland didn't begin until the 1730s.

What makes the larch so special is that it is Europe's only native deciduous conifer. It may look a little scruffy and stark in the winter, and its dullish-green appearance in midsummer is of no account, but in spring its vivid emerald-green needles and deep pink female flowers make it stand out amongst the other conifers. In the autumn the tree turns the most glorious golden yellow. Its virtue to the forester was perfectly summed up by Jacob George

Strutt in 1830: 'It is of quick growth, and flourishes best in poor soils and exposed situations, which renders it valuable in those places, where land is of little other value than to afford footing for such hardy mountaineers.' In fact, the larch has been so successful in Britain that it may be regarded as a naturalized species.

The oldest-surviving larch in Britain is a specimen in the Borders of Scotland, on the Kailzie Estate near Peebles. It was planted in 1725 and there is ample documentation of this date. Sir James Naesmyth, laird of Posso and Dawyck, offered one of the young larch trees that he had acquired in London to his good friend the Laird of Kailzie, who planted it in his garden. It thrives to this day, as does another specimen that Sir James took home to Dawyck.

More famous are the larches of Dunkeld. Strutt sketched the pair of handsome trees hard by the abbey in 1825, almost 90 years after their planting. A curious anecdote is told of their arrival on the estate of the Duke of Atholl in 1737 (or 1738, accounts vary). Colonel John Menzies of Megeny in Glen Lyon brought some young larch plants over from the Tyrol. He presented five of these to James, 2nd Duke of Atholl, 'along with orange trees and other Italian exotics, and they were all placed in the hot-house together; the temperature of the place, of course, speedily killed the Larches, and their remains were thrown on the dung-heap; here, their roots being covered by the refuse of the garden, some slight remains of life remaining still in them, they began to vegetate, and being in a more genial atmosphere, the branches shot forth their buds, and, by degrees, the plants became vigorous.' Their miraculous survival would lead to one of the most impressive programmes of tree planting in history. Within a few years it became obvious to the 2nd Duke that larch was admirably suited to the rugged upland terrain of his estate, for it produced better-quality timber than the native Scots pine and grew much faster. The 3rd Duke

Towering larches. The larch is a remarkably valuable and reliable commercial tree. After its introduction from Europe in 1725 it would change the whole face of British forestry, particularly in Scotland.

and, subsequently his son John, the 4th Duke (often known as 'The Planting Duke'), stepped up the cultivation of larch until, by 1830, the dukes of Atholl, among them, had planted more than 14 million larches, covering 10,000 acres. By 1812 the Dunkeld trees had become known as the 'Parent Larches' because they had provided so much seed for the earliest plantings. One of the original trees died when it was struck by lightning in 1909 but the other 'Parent Larch' or 'Mother Larch' is still alive and is one of Britain's largest, with a girth of 18 feet 6 inches.

In 2009 matters began to look rather bleak for the larch in Britain when a fungal pathogen called *Phytophthora ramorum* was found to have infected some Japanese larches (*Larix kaempferi*) in southwest England. Within a couple of years the pathogen had spread through Wales and on up the west side of Britain, killing thousands of hectares of commercial larch forestry. While Japanese larch is by far the most susceptible species, European larch and the hybrid Dunkeld larch (*Larix* x *marschlinsii*) are also at risk of infection. It would certainly be a tragedy if the 'Mother Larch' was among the fatalities, and one wonders if there may be a way of treating this unique heritage tree with some kind of fungicide should it ever develop symptoms. Ultimately there is every likelihood that the vast majority of larches in Britain will be killed by this deadly organism.

Larch timber is extremely durable and resilient and it is especially suitable for outside use, as it maintains its integrity in both wet and dry conditions. It makes an ideal nurse tree for protecting other species and encouraging competitive growth. Due to its relative lack of combustibility larch is also useful as a partial firebreak. The timber has been used for all manner of structural purposes, sometimes in preference to oak. In the early nineteenth century the 4th Duke of Atholl commissioned the building of a fine brig, constructed from larch timber, which he named (not surprisingly) the Larch. Between 1816 and 1820 a frigate of 28 guns was laid down at Woolwich, the first British warship to be built entirely of larch; named the *Athole*, she was built from timber supplied by the Atholl Estate. There were several advantages to using larch for marine purposes. Unlike oak, it did not corrode iron bolts and fittings, and as it was less prone to shrinkage it retained the oakum between the boards and did not need so much caulking maintenance. For warships in combat, the fact that larch is not prone to splintering caused far fewer casualties under cannon fire.

THE CEDAR OF LEBANON

The documentation of early tree introductions is often contradictory and this is certainly the case with the monolithic cedar of Lebanon (*Cedrus libani*). The tree clearly hails from the country that gave it its name and its image on the Lebanese flag is proof of its importance there as a national emblem. There are about 400 ancient trees still growing on Mount Lebanon and, with girths of up to 48 feet, many are believed to be as much as 2500 years old. The species is also widespread in Syria and in the Taurus Mountains of Turkey.

Popular opinion holds that Dr Edward Pococke, an English orientalist and biblical scholar, brought cedar of Lebanon seeds back to Britain from Syria in 1640. He was given the living at Childrey in Oxfordshire by Christ Church, Oxford, and he moved into the rectory there in 1642. It is alleged that he planted a cedar in his garden in 1646, although it seems strange that he would have held on to the seed for six years before deciding to sow it. This tree survives and is in good heart, making it (probably) the oldest cedar in Britain.

Cedars of Lebanon growing in the Chelsea Physic Garden in London as depicted by The Revd C. A. Johns in The Forest Trees of Britain. *They were reputedly planted in 1683, but sadly are now long gone.*

This vast cedar of Lebanon near the remains of Garnstone Castle in Herefordshire must be one of the biggest and widest-spreading examples in the country. The shape, with its distinctive layers and flat top, is typical of the species.

The cedar of Lebanon is rightly known as a giant among trees. Some are as much as 131 feet high and may achieve a canopy width of similar dimensions. Its massive presence was unmatched in its native range and it is frequently mentioned in the Bible, revered within the texts for its beauty, grandeur and strength. Its timber was valued by the ancient civilizations of Mesopotamia and Egypt as long as 4000 years ago for construction purposes and, of course, King Solomon used it for his great temple and his own palace at Jerusalem. The wood has always been prized not just for its strength but also for its delicate pink colour and its aromatic property, which was said to deter insect attacks. The ancients would collect the oil of the tree and treat papyrus documents with it to preserve them. Cabinet-makers have often used the wood to line drawers and chests, as it smells sweet, but whether it keeps the moths away is debatable.

It wasn't until the mid-eighteenth century that many of the cedars first planted in Britain began to bear cones and produce viable seed and when this happened it sparked a mania among the gentry. For the next hundred years it was the height of fashion to have these trees adorning the parklands and pleasure grounds of all the great houses of the land. What many of the later Victorian garden enthusiasts failed to appreciate, however, was that their cedars would eventually outgrow their modest plots. It is a pattern that would later be repeated with the craze for monkey puzzles. One major drawback for the cedar of Lebanon is its vulnerability to storm damage and the gales of 1987 and 1990 certainly took their toll on many historic specimens in the southern counties of Britain.

THE OTHER CEDARS

There are four true cedars: the cedar of Lebanon, the Cyprian cedar, the Atlas cedar and the deodar cedar. The Cyprian cedar

(*Cedrus brevifolia*) is native to Cyprus and Greece. A less statuesque tree and seldom grown in Britain, it is thought by some to be a subspecies or variant of the cedar of Lebanon. The Atlas cedar (*Cedrus atlantica*) is from the Atlas Mountains of Morocco and Algeria. The deodar cedar (*Cedrus deodara*) is a Himalayan species. It is not always easy to tell the cedars apart, particularly from a distance. In fact many early botanists, Sir Joseph Hooker amongst them, thought these trees were all slight variations of one species, representing various fragments of a once widespread cedar forest from before the ice ages. The mnemonic LAD is a useful aid to identity: L – for the Lebanon with its Level branches, A – for the Atlas with Ascending branches, and D – for the Deodar with Descending or Drooping branches. Often the boughs of large mature Atlas cedars will tend to level out, making it harder to distinguish from the cedar of Lebanon.

The deodar cedar is valued in India for all manner of construction purposes, for its aromatic timber is durable and highly resistant to insect attack, particularly by termites. Sacred to the Hindus, who use it to build their temples, it is called Devadera or the Tree of God. It is often burnt for its fragrant incense. Deodar seed first arrived in Britain in 1831, introduced by the Hon. Leslie-Melville, who planted it on the Melville estate in Fife. There is a tree grown from the original seed at Westonbirt House in Gloucestershire.

The Atlas cedar was discovered in 1827 and first introduced in 1839. In 1841, Lord Somers returned to Herefordshire with seed gathered at Teniet el Haad in Morocco and he planted Britain's first blue Atlas cedars (*glauca*) on his estate at Eastnor Castle; some of those original plantings still survive in the splendid gardens there. If anything, the blue Atlas cedar soon turned out to be the most admired and more frequently planted. None of the cedars produces timber of sufficient quality for major construction purposes in Britain.

THE MONKEY PUZZLE

Like the cedar of Lebanon, the monkey puzzle or Chile pine (*Araucaria araucana*) became the ultimate status symbol soon after its arrival in Britain in 1795. It was brought back from Chile by the Scottish physician and naturalist Archibald Menzies, who was commissioned by the English botanist Sir Joseph Banks to collect plants in the Americas. Menzies's first expedition was hugely successful but his second voyage, as ship's surgeon on board the *Discovery*, saw his plant-collecting activities curtailed by Captain Vancouver. The only new specimens he managed to bring back from his long trip along the western coast of the Americas were two strange little spiky plants grown from seeds that he had taken from a bowl of nuts served for dessert by the Spanish viceroy in Valparaiso. These nutritious nuts were the staple diet of the Araucana Indians of Chile, hence the formal name. The popular legend is that the tree's common name arose because someone somewhere once said that 'a monkey would be puzzled to climb it' but it is more likely that it comes from its association with the Pehuenche people of the south-central region of Chile, whose name translates as 'people of the monkey puzzle'. Native monkey puzzle forests here are currently under threat from over zealous and sometimes illegal logging. The monkey puzzle often looks out of place, if not downright bizarre, growing in many British settings, but it must have seemed like some scaly dinosaur when it was first introduced. It was its novelty value and exotic provenance that made it a firm favourite and focal point for planting about the grand houses and within the burgeoning public parks. They are clearly more suited to a larger landscape and one of the best places to see them at their most impressive is the long, straight drive leading up to Bicton House in Devon. The monkey puzzle avenue there was planted

One of the earliest plantings of monkey puzzles in Britain is this fine avenue at Bicton House in Devon.

in 1843 from seed brought back from Chile in 1841 by William Lobb, who worked for Veitch's Nursery in Exeter. Although the tree had been introduced 50 years previously, it wasn't until Lobb obtained large quantities of seeds, which Veitch could then market, that its popularity soared.

Close inspection of the tree's bark reveals a texture rather like the tough hide of a rhinoceros, which, in mature trees, may be as thick as 7 inches. This protects the tree from extreme climatic conditions in its native range, as well as resisting the intense heat of volcanic eruptions. Its primeval appearance is no coincidence, for trees of the same family are recorded in fossil deposits from the Jurassic period, some 225 million years ago, and 190 million years ago it was one of the dominant species of the southern hemisphere. Weird though it may seem, millions of years ago such trees once grew in northeast Britain; the dense, black material known as Whitby Jet, which Victorian craftsmen turned into jewellery, is actually the highly compressed fossilized wood of *Araucaria*.

DAVID DOUGLAS AND HIS INTRODUCTIONS

Menzies first discovered what would later become known as the Douglas fir at Nootka Sound in British Columbia in 1792, though it would be another 35 years before it was introduced to Britain. The man responsible was David Douglas, who was born in 1799 on the Scone estate near Perth. As a young boy Douglas developed a keen interest in natural history and by the age of 11 he was serving as an apprentice gardener on the Scone estate. He showed academic flair, with a prodigious thirst for botanical knowledge, and by 1820 he had secured a post at Glasgow's Botanic Gardens. Here he worked under the great William Hooker, and when Hooker was approached by the London Horticultural Society (founded in 1804) to recommend a botanical collector who might work for them, he immediately suggested Douglas. From 1823 Douglas made several trips to America, beginning with a visit to the east coast. Over the next 11 years he made progressively more arduous and exhausting forays into the most rugged of territories in the Pacific northwest. Douglas's journals provide detailed accounts of his epic journeys and his botanical discoveries. He often put his life in danger, surviving foul weather, capsizing in rapids, falling down ravines, clashing with Native Americans and almost starving to death several times in his quest for new species. He was 35 when he accidentally fell into a wild bull pit in Hawaii and was gored to death. In all he was responsible for introducing 254 new plants and trees to Britain but his name will be for ever associated with the mighty conifers that he shipped back, for these were destined to change the nature of British forestry completely.

A portrait of David Douglas (1799–1834)

The Douglas fir (*Pseudotsuga menziesii*), whose popular and Latin names combine to pay tribute to both Douglas and Menzies, is now one of Britain's most important conifers. Douglas explored the lower reaches of the Columbia river in 1825 and brought the seed to Britain in 1827. The first seedlings were produced in the nursery at Scone and one of that first batch was planted in the garden there in 1834, the year that Douglas died. It thrives today as an imposing specimen tree and a living memorial to a great man. The tallest Douglas fir in Britain is in Reelig Glen in the Scottish Highlands, a very respectable 217 feet. This is not as impressive as it sounds, for the tallest Douglas firs in America are more than 330 feet high. The big advantage to foresters is the speed at which these trees grow, often as much as 98.4 feet in 30 years, and they will outstrip virtually all the competition.

Douglas discovered two other conifers in the Columbia River region in 1825 – the grand fir (*Abies grandis*) and the noble fir (*Abies procera*). By the end of the nineteenth century the grand fir had proved itself to be an excellent timber tree in Britain, particularly in Scotland. The noble fir, on the other hand, was found to have less potential as a commercial crop, though it makes an attractive ornamental and does well in very moist and poor acidic soils. In 1826, Douglas found the western yellow pine (*Pinus ponderosa*), also known as the blackjack or ponderosa pine, growing in the Spokane River region. It gets its common name from the yellowish hue of its bark and is one of western America's most widespread and important timber trees. The Monterey pine (*Pinus radiata*) was found on the Monterey Peninsula in California and it was first planted in Britain in the Chiswick gardens of the London Horticultural Society in 1832. Because it comes from a warmer climate it has done best along the south coast, particularly in Devon, where large, old, gnarled, wind-shattered trees face out to sea. As a timber-producing tree it is very important in Australia, New Zealand, South Africa and parts of South America.

The single most important tree for commercial forestry in Britain that Douglas introduced is probably the Sitka spruce (*Picea sitchensis*). Again it was Menzies who first discovered this tree, in Washington State in 1792, and Douglas who shipped the seed back to Britain (almost 40 years later). Recognition of the tree's commercial potential was somewhat tardy, for it was not until the formation of the Forestry Commission in 1919 that trials established that this was a remarkably adaptable tree on almost any terrain, capable of producing consistently high yields of strong, lightweight timber, which made it ideal for pulping to make paper.

THE REDWOODS

The massive trees of America known collectively as the redwoods have been around for millions of years and fossil remains indicate that their range was widespread across the northern hemisphere. Evidence of them has been found in Cornwall and the Isle of Wight. Now the coast redwoods (*Sequoia sempervirens*) and their cousins the Wellingtonias or Sierra redwoods (*Sequoiadendron giganteum*), both members of the swamp cypress family, have retreated to two strongholds in the southwestern United States. From a distance the trees do look similar but the foliage is quite different. The coast redwood has two types of foliage – needles that are a bit like those of the yew and, to a lesser extent, small-scale leaves on leading shoots – while the Wellingtonia has long fingers of tough, pointed scales, much like a cypress. Coast redwoods are found in a narrow coastal strip from California up to Oregon, and the Wellingtonias grow on the western slopes of the Sierra Nevada, some 200 miles to the east. The two species never mix in their native ranges. Loggers took a terrible toll on the redwoods in the nineteenth century and commercial timber companies continued to exploit them until quite recently. Only the establishment of national and state parks and the tireless energy of redwood conservation bodies have managed to protect the last remnants of these once vast forests.

The coast redwood holds the record as the tallest tree in the world. When tree writer and photographer Thomas Pakenham visited California in the 1990s he saw the Stratosphere Giant, then an impressive 368 feet 7 inches high. He learnt that there were at least 25 other trees in excess of 360 feet, so the champion's claim is not secure. The largest-ever cut stump to be measured gave a count of 2200 annual rings. Coast redwoods have a remarkable feature that no other conifers possess: they are able to regenerate from a cut stump, much like a broadleaf tree will do after coppicing. Born survivors, these trees can also grow from small burls, knobbly growths containing dormant bud material, which fall from the trees and then put down roots. When trees fall in the forest the broken boughs that hit the ground often develop new roots and branches on the top side of the bole may form a row of new trees, which may eventually go on to grow independently. There is a remarkable example of this in Leighton Grove on the Welsh Borders, where a redwood that was blown down in 1936 is still alive.

The first European to report back on the coast redwoods was a Franciscan missionary, Father Juan Crespi, who travelled with the Spanish Portola Expedition of 1769. Exploring the area around Monterey Bay in California, he noted that it was 'well forested with very high trees of a red colour, not known to us'. They looked a bit like cedars but Crespe realized that he had come across something quite different and it was he who gave them their common name. Menzies was one of the scientists on the *Discovery* expedition

This prostrate coast redwood in Leighton Grove, near Welshpool in Powys, was blown down in 1936. Foresters decided to leave it alone and it soon put down roots from the broken boughs that hit the soft woodland earth. Several large trees now grow along the old trunk, mimicking the typical mode of regeneration found in the forests of the tree's native territory along the coastal region of California.

in 1794 and he collected herbarium specimens in Santa Cruz. In a letter to Douglas he claims to have rediscovered the species in 1832, again collecting seeds and plant specimens. Mysteriously, none of these finds ever made it back to Britain and it was 1843 before the first seed arrived, sent by physician and botanist Dr Fischer to the Physic Garden at Chelsea. In Britain the largest coast redwoods are a mere 131 feet tall, but then they are relative novices compared to the record-breakers of California, which may be at least 500 years old.

One particular feature that distinguishes the Wellingtonia from the coast redwood is its phenomenal splayed and buttressed base. Some of the American specimens are estimated to be around 4000 years old. There are two old veterans known respectively as General Sherman and General Grant and they have been rivals since they were first discovered in the Sierra Nevada mountains of California in the mid-nineteenth century. General Sherman has proved to be the winner at an astonishing 55,040 cubic feet, making it the biggest living thing in the world, compared to General Grant at 47,930 cubic feet. Modern statisticians now judge a tree's size on the basis of volume of timber rather than girth and height.

There is some controversy over who brought the first Wellingtonia seeds to Britain. In July 1853, Scottish botanist and plant collector John Matthew sent a letter containing some seeds to his father in Perthshire and these were planted. However, this information was not published until June 1854. Meanwhile, William Lobb, who was based in San Francisco, had brought some seeds to Britain and the details appeared in the *Gardeners' Chronicle* of 24 December 1853. Lobb's employer, Veitch's Nursery, went on to produce Wellingtonia specimens for the discerning (and wealthy) Victorian plant enthusiast. They could not have wished for better publicity when John Lindley of the London Horticultural Society named the tree *Wellingtonia giganteum* in honour of the Duke of Wellington, who had died in 1852. After this there was scarcely a private estate or public park in the land that did not plant one, or in some cases a whole avenue, as a memorial to Britain's great hero. They became even more popular than the cedar and the monkey puzzle.

The Wellingtonias in Britain still have a lot more growing to do before they reach their full potential. Currently, the tallest Wellingtonia is at Castle Leod, Strathpeffer, in Easter Ross – at 170 feet it's only got another 100 feet to go before it matches General Sherman. Estate records show that the tree was planted in 1853 to commemorate the first birthday of Francis Mackenzie, Viscount Tarbat (later the Earl of Cromartie). This is one of the few known survivors from the original batch of introduced seed.

THE GINKGO – THE LIVING FOSSIL TREE

There is one tree that has a unique place in the natural order of plants – the ginkgo (*Ginkgo biloba*). Though closely related to the conifers, it has a different mode of fertilization: pollination occurs by means of self-propelling sperm, which seek out the female ovaries. This is a feature that makes it more akin to the palm-like cycads than a tree. Indeed, the ginkgo leaves closely resemble those of the maidenhair fern, which accounts for its alternative name, the maidenhair tree. In eastern China, where its native range is in Zhejian and Guizhou provinces, it is commonly called the duck's foot tree, which is a reference to its leaf shape. The name 'ginkgo' is a Japanese derivation of the Chinese yin-kuo, which means 'silver fruit'. The trees can grow to a prodigious size, achieving heights of 131 feet and boles of 13 feet in diameter. Many have been growing for around 1000 years and it is estimated that some may be as much as 3000 years old.

Often known as the 'dinosaur tree', the ginkgo and its relatives had a worldwide distribution between 150 and 200 million years ago, long before dinosaurs first trampled through them. The very idea that any living thing could have existed for this length of time, virtually unchanged in form, is almost beyond the bounds of the imagination, yet fossil deposits do bear this out. Fossils have been found in several parts of Britain, commonly around the northeast coast, and the distinctive leaves resembling a fan, butterfly or duck's foot, with their central notch and lack of the central rib normally found on broadleaves, clearly mark them out. As with many other exotic species, it was the effects of successive glaciations that saw their demise in Europe around three million years ago. Botanist and arboriculturist Tony Kirkham describes this reshaping of the world's forests most eloquently in his book *Plants from the Edge of*

The 'Old Lion' ginkgo in Kew Gardens was one of the first trees to be planted in the gardens in 1762. It was grown in the Mile End Nursery in 1754 from the first seed brought from China.

the World. However, he also points out that in the wake of these colossal climatic changes over millions of years Britain once again has an ideal climate in which all these long-lost trees can thrive. Due to the enthusiastic endeavours of plant hunters and collectors over the last few hundred years many 'lost' species of trees have returned to their ancient home, contributing to one of the most diverse tree populations in the world.

The German physician and botanist Engelbert Kaempfer was the first European to discover the ginkgo. When he visited Japan in 1690 he observed that the Buddhist monks venerated this tree and often planted them around their temples. He published his findings in his *Amoenitatum Exoticum* in 1712. It was probably shortly after this that the ginkgo arrived in Europe; the first seed to be planted in Britain was at a nursery in the Mile End district of London in 1754. The Duke of Argyll bought one of their specimens for his estate at Twickenham and shortly after he died the tree was carefully lifted and taken down the Thames by barge to Kew. Lord Bute saw the tree safely transplanted in 1762 as one of the inaugural specimen trees of the Botanic Garden, where it flourishes today as one of Kew's 'old lions'.

Even though it may not be correct to think of the ginkgo as a conifer, the form of a mature tree is not dissimilar to the larch, with a range of foliage colour throughout the seasons to rival the very best of the deciduous trees: a vivid yellow-green in spring and then, as autumn beckons, a more delicate tinge of yellow, that slowly suffuses inwards from the leaf edges until the whole tree has turned a deep gold. The male trees are usually preferred for cultivation, as the fruits of the female tree tend to make a squishy mess when they fall and smell unpleasant as they rot. The nutritious nuts are often roasted and eaten in China and are said to counteract the effects of over-indulgence of alcohol. They have medicinal virtues too and are used to alleviate asthma and bronchitis as well as digestive and urinary problems. An extract from the dried leaves is also used as a diet supplement and aids a variety of blood disorders. There are claims that it can help to improve the memory and may prove useful in the treatment of Alzheimer's disease. Northern parts of Britain are a little chilly for the ginkgo but it has been frequently planted in parks and gardens of southern England. In America its true potential as a street tree has been realized, as it is particularly resilient in a polluted atmosphere. It is also remarkably resistant to pests and diseases, for those which might have harmed such a tree appear to have died out with the dinosaurs. It is certainly a tree well worth considering in the future for urban Britain.

THE DAWN REDWOOD

In 1941 two remarkable coincidences occurred that led to the stunning revelation that a tree from a genus previously known only from fossil records was actually very much alive and well in a remote Chinese valley. Professor Shigeru Miki, studying fossils found in Japan from the Pliocene period (1.5–6 million years ago), realized that, although very similar to the redwoods and swamp cypress, he had something quite different before him, for here the leaf shoots and the needles were in opposite pairs rather than alternately arranged. He called his new find *metasequoia*, meaning 'like a sequoia'.

Matters might have rested there had it not been for the simultaneous discovery of another new tree, some 3000 miles away on the Sichuan borders of China. The people of Mo-tao-chi village, who were in the habit of using the foliage of this tree as fodder for their cattle, called it a water fir. The forester who had come across it sent samples to the Central Bureau of Forest Research, who passed them on to Professor W. Cheng at Chungking University. The experts were stumped. They thought it looked very similar to the Chinese swamp cypress (*Glyptostrobus pensilis*), yet it was definitely different. Time rolled on and Professors Hu Ziansu and Zhęng Wanjun of the Fan Memorial Institute of Biology in Beijing finally saw specimens from the tree in 1946. They brought all the known information together and reached the conclusion that this was indeed Professor Miki's fossil tree. It was given the name *Metasequoia glyptostroboides*, known popularly as the dawn redwood. Not long after, several dawn redwoods were discovered growing in a valley about 30 miles away from Mo-tao-chi village.

By 1947 the Chinese were corresponding with botanists in America, establishing an east-west interchange that permitted seed to be sent to the west, initially to be distributed by the renowned Arnold Arboretum. The dawn redwood has since flourished in all temperate regions on all types of soil. Like some ancient jewel released from a time capsule it has burst back into life across so much of its primeval range. The first tree to be planted in Britain was in 1949 at the botanic gardens at Cambridge University. A tall, conical tree with a light and airy feel, it is one of the few deciduous conifers and it makes a superb ornamental species. In spring it has vivid green foliage, delicate and soft to the touch, changing to a warm pinkish orange in the autumn. The tallest dawn redwood in Britain was planted around 1960 in Wayford Woods in Somerset and has already reached 105 feet.

THE WOLLEMI PINE

If the discovery of the dawn redwood were not fabulous enough, then imagine the thrill that ran through the botanical world after a very similar chance discovery was made in Australia in 1994.

David Noble, a national parks and wildlife officer in New South Wales, was out bushwalking in a remote part of the Wollemi National Park in the Blue Mountains when he came upon a rather unusual conifer growing on moist ledges, tucked away in a deep rainforest gorge. Having never seen one before, he was intrigued and he took away a small fallen branch to see if he could identify it. He drew a blank, but sought expert advice. The outcome was to stun everyone with an interest in the natural world, for Noble had discovered another so-called dinosaur tree.

The specimen he had brought back was identified as a member of the Araucariaceae family but of a genus known only in fossil deposits, the oldest of which were some 90 million years old. Since the dinosaurs were around until 65 million years ago, it's quite feasible that they dined upon this tree, which was given the name Wollemi pine (*Wollemia nobilis*). Although not strictly a pine tree it is a close relative of other Araucarias with the pine appellation, namely the Kauri pine, the Norfolk Island pine, the Hoop pine, the Bunya pine and the monkey puzzle. It was thought that Wollemi pine had died out at least two million years ago, yet here it was, less than 130 miles west of Sydney. It is amazing that no one had spotted it before but Noble's discovery rewarded him with something that all naturalists crave (if they're honest) – a species named after himself.

Only about 100 mature specimens have been found in the wild so far, although the biggest of these is a respectable 131 feet and 3 feet 11 inches in girth. It is thought that this individual may be as old as 1000 years. The tree has very distinctive, pendulous, dark green foliage and a strange, bubbly textured, brown bark, which looks a bit like masses of Kellogg's Coco Pops. It has a growth habit of throwing multiple stems from the base in what has been described as self-coppicing – unusual in any conifer.

A comprehensive conservation and propagation programme has been set in motion to guarantee the future of the Wollemi pine and the intention is to distribute it as widely as possible so that its survival is assured across a range of different countries and environments. It seems to be a remarkably adaptable tree so far, growing in a wide range of temperatures and situations, and even doing well as a pot plant. The funds raised from a concerted marketing initiative will be reinvested in the conservation programme for the tree in the wild.

Wollemi is an aboriginal word meaning 'look around you, keep your eyes open and watch out'. In so many respects this epitomizes the ardent endeavours of the plant hunters and collectors down the centuries and this should also be the maxim for those seeking the next Wollemi or dawn redwood. The natural world still has many more secrets up its sleeve.

After the first edition of this book there were one or two dissenting voices criticising the tree's inclusion in a book entitled *The Trees that made Britain*. How on earth could this johnny-come-lately be seen to have made an impact on the British treescape after such a short time? Taking this on board I will now qualify it as a tree that is 'making' Britain's future treescape, for it has been widely planted in parks, gardens and arboreta and so far it is doing rather well.

Orchards

Orchards are some of the most special places in the British countryside, so why do so many landscape historians and tree and woodland conservationists ignore them? They are much more than a mere extension of the domestic garden or a farmer's field. Sadly, the Oxford English Dictionary's definition of an orchard as an 'enclosed piece of land with fruit-trees' applies as much to modern intensive cultivation as it does to the real deal.

The wind among the apple-trees
Is murmurous as distant seas,
And from my orchard-shaded house
I see the sea-green of the boughs
Break into crests of blossomy spray,
Where the adventurous bees all day
Ride on the surf, and butterflies
Spread fragile canvas to the skies,
Yacht-like to sail my orchard seas.

FROM 'ORCHARD IDYLL' BY SYDNEY MATTHEWMAN,
TAKEN FROM 1947 ANTHOLOGY SPIRIT OF THE TREES

A BOOKE OF THE
Arte and manner how to Plant and
Graffe all ſorts of Trees, how to ſet
Stones and ſow Pepins, to make wilde Trees
to graffe on, as alſo remedies and medicines.
With diuers other new practiſes, by one of the Abbey of S. Vincent in France, practiſed with his own hands,
deuided into vij. Chapters, as heereafter more plaine-
ly ſhall appeare, with an addition in the ende
of this booke, of certaine Dutch practiſes, ſet foorth and Engliſhed by
Leonard Maſcall.

In laudem incisionis distichon,
Heſperidum Campi quicquid Romanaque tellus,
Fructificat nobis, incisione datur.

¶Imprinted at London by T. Eſte,
for Thomas Wight. 1599.

An early woodcut of the title page of Leonard Mascall's 1599 guide to the planting and management of orchards, with a wonderful picture of a gentleman engaged in top-grafting his trees.

Old-fashioned orchards tend to be diverse, combining apples and pears with cherries or plums, surrounded by hedges of damson, hazel and elder. Fruit trees can survive for 100 years or more and some pear varieties will live for 300 years. As they age they will crop less well and become misshapen, attracting mosses, lichens and mistletoe, which they tolerate quite happily. But what they lack in productivity they more than make up for in providing a superb wildlife habitat, rich in invertebrates, and a marvellous source of food and nesting sites for birdlife. And if the vegetation between the trees is left untouched, grasses, fescues and wild flowers will flourish.

THE RISE OF ORCHARD CULTURE

There have been wild fruit trees native to Britain for many thousands of years and it is known from studies of middens associated with prehistoric settlements that crab-apples, wild pears, wild cherries, sloes and hazelnuts formed part of the human diet. There is evidence that there was some kind of cultivation of fruit trees in Turkey and the Caucasus as long as 5000 years ago. The Greeks and the Romans took to the whole idea of fruit culture with great enthusiasm, and it was the horticulturally erudite Romans who brought their knowledge of orchard culture to Britain, introducing apples, pears and plums as well walnuts and sweet chestnuts.

After the Romans left in the fifth century little is known about orcharding until the late Middle Ages, when records show that the monastery garden was the place where fruit and vegetable cultivation was most widely practised, with new ideas about horticulture spreading among the religious orders all over Europe. The very earliest image of apple trees in a monastery garden, at Christ Church, Canterbury, dates from around 1165.

Commercial orchards started to develop in the mid-sixteenth century. In 1533, Henry VIII commissioned Richard Harris, then the royal fruiterer, to plant extensive experimental orchards across 140 acres near Teynham in Kent, using graft wood imported from France and Holland. There is a reference to this in *The Fruiterer's Secrets* of 1604:

> The Dutch and French, finding it [fruit] to be so scarce, especially in these counties neere London, commonly plyed Billingsgate and divers other places with suche kinde of fruite. But now (thanks to God) divers Gentlemen and others taking delight in the grafting have planted many orchards, fetching these grafts out of that orchard which Harris planted. And by reason of the great increase that now is growing in divers parts of this Land of such fine fruit, there is no need of any foreign fruit, but we are able to serve other places.

During the seventeenth century books about the planting and management of gardens began to appear and these often included detailed instructions on how to establish fruit trees and orchards. One of the most important of these early works is *Paradisi in Sole Paradisus Terrestris* by John Parkinson. He was apothecary to King Charles I, who bestowed the title Botanicus Regius Primarius upon him when the book was published in 1629. Parkinson owned a fine

A fine, old, traditional apple orchard in flower, with rows of trees on standard root stocks.

garden in Long Acre in London, which, it is said, contained 484 different types of plant. Available varieties of apples were clearly increasing at this time, for Parkinson lists 57 sorts. Here at last was a manual that gardeners could use to learn about the finer points of orchard culture. In the 1650s, Samuel Hartlib and Ralph Austen both added new works in support of orcharding, Hartlib now alluding to more than 200 different sorts of apple. John Evelyn, who moved into his house at Sayes Court in Deptford in 1652, must have been familiar with these books and he reinforced the drive for more orchards in his own book, *Sylva*. His superb plan of Sayes Court shows: 'The great orchard planted with 300 fruit trees of the best sorts mingled and warranted for 3 years upon a bond of 20 pounds.' Evelyn was clearly a shrewd man, covering himself against failure of his new fruit trees.

In the late seventeenth century the first dwarf and semi-dwarf rootstocks were introduced, which meant that people with quite modest gardens were now able to find space for fruit trees. They could be trained into espaliers, formed into hedges or walks and even grown in pots. They were easy to manage and easy to harvest. Near his great orchard Evelyn had 'a palisadoed hedge of codlins and pearmaines'.

A saviour for orcharding emerged in the late eighteenth century in the shape of Thomas Andrew Knight of Downton in Herefordshire. He had seen how Britain's fruit trade was suffering from competition from abroad and knew that many of the varieties were poor croppers prone to disease. He made it his mission to develop new, improved, superior-cropping, disease-resistant varieties. His meticulous hybridizing experiments produced not only many fine new apples but also a whole range of pears, cherries, damsons, nectarines and strawberries. He also turned his attention to potatoes, peas and cabbages. He established trial gardens

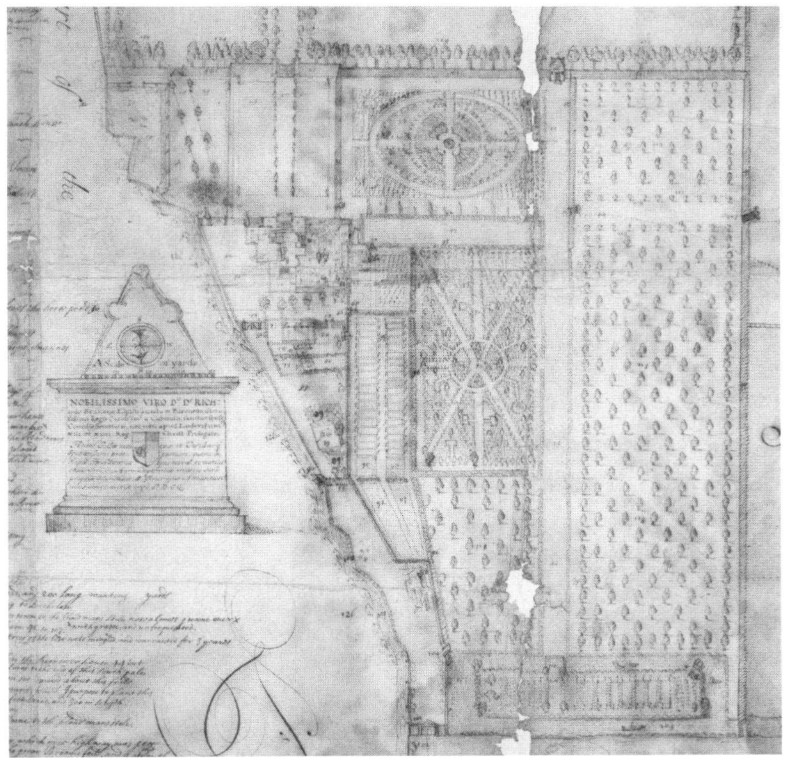

The original plan of John Evelyn's gardens at Sayes Court, Deptford, from the 1650s. Note the large regularly-spaced orchard on the right.

where more than 1400 fruit specimens were grown and monitored, and published his *Treatise on the Culture of the Apple and Pear* in 1797, which subsequently went to four editions. Knight was highly regarded by his peers in the London Horticultural Society, becoming its second serving president in 1811, the year that he produced his *Pomona Herefordiensis*, illustrated with 30 superb hand-coloured engravings of fruit.

Knight's influence was dramatic, for his well-publicized research encouraged plantsmen and landowners all over Britain to plant the new varieties he had developed. There seems to have been something of a dip in the fortunes of orchards in the mid-nineteenth century but in the latter decades, with improved transport and communications leading to better market supply, orchard fruits once more began to prosper. Interest in apples spawned the British Pomological Society, which was founded in 1854 to promote and advance apple cultivation, and the Royal Horticultural Society moved in on the act in 1883 with its Great Apple Show, attracting 183 exhibitors. Botanist G. S. Boulger, writing in the 1880s, stated that there were then 5000 varieties in cultivation. However, it's likely that this is an inflated figure as the same varieties often had more than one common name.

The first half of the twentieth century was a period of prosperity and growth for the orchard industry, with more than 273,000 acres given over to fruit cultivation by the early 1950s. Unfortunately, the second half of the century saw a steep decline as agricultural fashions changed. Cheap fruit flooded in all year round from every corner of the globe and British producers found it hard to compete. Thousands of orchards were grubbed out in favour of arable crops and many of the smaller plots on the urban fringes were sold off to property developers. By 1971 there were barely 56,000 acres left – a decline of almost 80 per cent. As a consequence Britain has suffered a reduction in the national fruit gene bank.

For a while the traditional orchard looked like a lost cause. Then in 1983 a small group of people set up an organization called Common Ground. They really cared about the British landscape and local distinctiveness and they have done their utmost to save the traditional fruit varieties from extinction by campaigning to save what was left of the old orchards and encouraging the planting of new ones. Together with other bodies, such as the Countryside Agency, English Nature and the Woodland Trust, Common Ground raised public awareness of the whole subject, as well lobbying both local and national government. As a result, authorities now realize how important such places are to their communities and conservationists have a better understanding of their rich habitat significance. Even the consciences of the bigger supermarkets have been pricked (occasionally) and some of them are now more inclined to stock a few of the more obscure English apple varieties, nudging aside the tasteless Golden Delicious and Granny Smith for a brief season.

APPLES AND CIDER

In spite of the limited range available in the average British supermarket, there are still a lot of apple varieties to be found if you know where to look. Manfully hanging on to the nation's heritage is the Brogdale Collections in Kent. This national archive of all things fruity and nutty maintains a stunning collection that includes over 2200 different varieties of apples, 550 pears, 337 plums and 285 cherries, not to mention the quinces, medlars and nuts. A large proportion of these fruits will have been imported varieties at some time in the past, but of the apples around 750 are thought to be truly British. There are also many places all over the world that have adopted them, notably the east coast of America, where the early settlers took their favourite varieties with them. Some of these old and long-forgotten apples have now returned home.

Most cultivated apples fall into one of three distinct categories and a few are multi-purpose. Dessert apples are sweet and ideal for eating. Cooking apples are tart and usually need sugar to make them taste sweeter. Cider apples, often known as vintage fruit, are generally too acidic and high in tannins to be palatable. In addition, the sour-tasting wild crab-apples make excellent wines and preserves.

Early records of apple varieties seem rather vague before the seventeenth century and they were often referred to as general types. Although there is little hard evidence, it is generally believed that the Court Pendu Plat and the Decio varieties, still available today, would have been cultivated by the Romans, and Decio was certainly known in Italy during the Middle Ages. Both the Pearmain and Costard apples are known to have been grown during the thirteenth century. The Pearmain, a variety that owes its name to the rather elongated form of the fruit, is mentioned in a deed of 1204 in Norfolk. The Costard first appeared on a fruiterer's bill to the court of Edward I in 1293 as 'Poma Costard', a name derived from the apple's shape – ribbed or costate. Costardmongers, who originally sold these apples, is a term little changed to this day, as these are the people who still sell fruit and indeed vegetables from street barrows. Costard was apparently medieval slang for 'head', suggesting that these must have been very large apples. Although a most important kitchen apple until the late seventeenth century, they are now unknown.

Archie Miles picking apples in his orchard at Hill House Farm.

There are many early references to codlins or codlings, which were the apples most favoured for apple sauce or pies as they tend to cook down to a soft, frothy consistency. The Keswick Codlin, first discovered around 1793 near Ulverstone in Cumbria, is a well-known type in many regions today. Various pippins have been around for hundreds of years and the name indicates that the variety is derived from a pip. Pippins are conferred upon the fruit-growing world all the time, often as a result of some apple core idly tossed away, but only a tiny proportion will ever amount to a variety that is worth cultivating. Golden Pippin is one of the first recorded and it seems to have been popular in the sixteenth and seventeenth centuries. One of the most important, and best documented, is the Ribston Pippin. Sir Henry Goodricke of Ribston Hall in Yorkshire obtained three pips during a visit to Rouen in France around 1688. He planted these at Ribston but only one survived. The Ribston Pippin, also known as the Glory of York, was a winner and many people obtained trees derived from this one original. However, the Ribston Pippin's fame and excellence didn't stop there for it became one of the parents to the now ubiquitous Cox's Orange Pippin, first raised from seed by Richard Cox of Colnbrook Lawn near Slough in 1825. The other parent of the 'Cox' was another famous pippin, the Blenheim Pippin, better known as the Blenheim Orange, but also called Woodstock Pippin (where it was first raised), Kempster's Pippin (by whom it was planted) and another 63 different names besides in various parts of Britain. Originally raised around 1740 on the edge of Blenheim Park in Oxfordshire, the Blenheim is now quite uncommon but it is a splendid apple, suitable for eating as well as cooking. Its excellent keeping quality gives it yet another name – the Christmas Apple.

Seedlings were also apples grown from seed and, as with pippins, every so often a wonderful apple is discovered. The most famous of these is the Bramley's Seedling. First planted between 1809 and 1813 by Mary Ann Brailsford in her cottage garden in Southwell in Nottinghamshire, the seedling bore its first fruit in 1837. In 1857, Matthew Bramley, who by then owned the cottage, invited renowned nurseryman Henry Merryweather to take cuttings of his tree on the understanding that any plants that were sold commercially should bear his name. The Bramley's Seedling had arrived, although it was really Brailsford's Seedling and would go on to be the world-beating cooking apple it is today, with a UK market alone estimated to be worth around £50 million a year. The original tree still exists in the modest cottage garden in Southwell, although sadly it is now in a sorry state as it is gradually dying due to an unstoppable attack from honey fungus.

The part that apples have played in British social and cultural history is well documented in the endless supply of tales of discovery, associations with time, place and personalities, as well as the historic catalogues of specialist nurseries such as Laxtons in Bedford. Apples were named not only in honour of kings and queens but also ordinary folk and farmers. We know who King Charles and Lord Kitchener were but what about Granny Giffard and Polly Prosser (from Kent) or Jennifer Wastie and Old Fred (from Oxfordshire). Individual farms or villages have lent their names to apples, as have the shapes and colours of the fruit, and many of them reflect the self-confidence of their growers – Wonder, Beauty, Favourite, Nonpareil, Excelsior, Superb, Prolific and Victory. If the plucking of an apple fresh from the tree to eat is a delight, and the smell of the season's first apple pie, fresh from the oven, is mouth-watering, then to sample the amber pleasure of the best cider completes the experience. In a discourse for the Royal Agricultural Society, Frederick Faulkner, writing in 1843, championed the production of cider on the farm: 'An orchard affords an agreeable variety in

The original Bramley's Seedling apple tree in Southwell back in 2002 with the late Nancy Harrison, the proud guardian of this special tree for many years.

Cider-making in Devonshire as depicted in the Illustrated London News *of 14 December 1850.*

the farmer's hopes and pursuits, and no inconsiderable addition to his domestic comforts and enjoyments. It is, indeed, the Englishman's proper and natural vineyard, producing him, almost without labour, fruit of rich and various flavour, more beautiful than the grape, and yielding an abundant supply of a scarcely less agreeable and cheering beverage.'

For too long we've been satisfied with the syrupy saccharin-laden commercial excuse for cider. When in cider country, be it Devon, Somerset or Herefordshire, it's worth seeking out the small cider producers and getting a taste of the real thing, and that means cider made from pressed apples rather than concentrates imported from the other side of the world. Cider in these counties has been common currency for hundreds of years; made on the farm it was customary to part pay the labourers with cider. It has always been considered an invaluable country craft to be a master cider maker. Today there is a steadily growing group of enthusiasts, with their orchards of fine old vintage fruit, who are once again reviving a tradition that, until recent years, had all but petered out. Good cider is being made from glorious concoctions of many different varieties, usually a blend of sweet, bittersweet and sharp apples, designated thus by their respective tannin and acid content. Great cider is the result of some select single-variety pressings, usually made from bittersweet varieties, at which point the drink is barely distinguishable from a fine wine. The Holy Grail for single-variety makers is the celebrated Kingston Black.

When the crop is all safely in and the cider is churning contentedly in the great vats, it is time, in midwinter, to ensure that a fine harvest will be garnered in the following season. On Twelfth Night in the western and southwestern counties of England there has been a long-standing custom of apple wassailing, a ceremony held to venerate the tree spirits. Wassailing has been traced back

'Wassailing apple-trees with hot cider in Devonshire on Twelfth Eve.'
from Illustrated London News *12 January 1861.*

as far as the fourteenth century but it may well have a much older origin. 'Waes Hal!' was an Anglo-Saxon greeting meaning 'Be Healthy' or 'Good Health' (still a popular salutation when taking a drink together). At the apple wassail the villagers would congregate in their orchard to bid 'Good Health' to their trees. This usually involved pouring libations of cider over the roots of the biggest tree in the orchard, wedging cake or toast soaked in cider in the fork of the tree as sustenance for the tree spirit and then making a great din with drums, pans and guns, as well as the setting of bonfires to awaken the good tree spirits and drive away the demons, thus ensuring a fine crop in the coming year. This would be accompanied by a good deal of singing and drinking from the communal wassail bowl, brimming with the previous year's cider, often warmed and mixed with spices, eggs or honey. There are many regional variations of the wassail rhyme or song but the following version is recalled by folklore expert Ralph Whitlock, who remembers it being sung in Norton Fitzwarren in Somerset:

Old apple tree, Old apple tree,
We Wassail thee, and hoping thou wilt bear,
For the Lord doth know where we shall be
Till apples come another year;
For to bear well, and to bloom well,
So merry let us be;
Let every man take off his hat and shout to thee,
Old apple tree, Old apple tree,
We Wassail thee and hoping thou wilt bear,
Hat fulls,
Cap fulls,
And three bushel bag fulls,
And a little heap under the stairs.

As many of the old, traditional orchards have disappeared and left only memories in many villages, the custom has slowly waned, but latterly just a few places are staging something of a revival. Usually set in motion by an enthusiastic group of local morris men, apple wassails are staged with all the elements one would expect, with the addition of plenty of music and song, mumming plays and even fire-eaters and jugglers. People come from far and wide on a bitter winter's eve to enjoy the celebrations.

PEARS AND PERRY

As I sit here writing, I take another draft of last year's perry and I briefly catch a whiff of the evocative scents and sounds of a late October afternoon, when I was bent double beneath a great old pear tree, bagging the ripe, fallen fruit. I left a few windfalls behind as a treat for the blackbirds and thrushes and then a few greedy sheep burst through the fence to hoover up the leftovers.

Outside Herefordshire, Worcestershire and Gloucestershire the mention of perry pears usually draws a blank response and yet for anyone brought up in these counties they are a most distinctive aspect of the landscape. Wild pears may occasionally crop up in woodland or hedgerow in various parts of the country but they are usually scruffy, undistinguished-looking trees.

Perry pears are thought to be either a hybrid of wild pears and introduced dessert varieties or they were specific, introduced types that arrived in Britain with the Normans in the eleventh century. There is evidence of a long tradition of perry-making in France going back to the fall of the Roman Empire. The drink has traditionally been used as a tonic to aid digestion and the Romans believed that it was the best antidote to the effects of poisonous mushrooms. The theologian Palladius, writing in the fourth century, claimed that the 'wine' made from pears (called castomoniale) was superior to that made from apples and he provided his readers with a recipe for it.

In 1580 the cleric and social historian William Harrison described 'pirrie' (the term comes from the Anglo-Saxon word

These beautiful perry pear trees at Holme Lacy, near Hereford, are actually the layered remnants of a single massive tree.

pirige, meaning 'pear') as a drink that the people of Sussex, Kent and Worcestershire made in addition to cider. The English botanist John Gerard, writing in 1597, was an enthusiastic advocate: 'Wine made of the juce of Peares, called in English Perry, is soluble, purgeth those that are not accustomed to drink thereof; notwithstanding it is as wholesome a drinke being taken in small quantities as wine; it comforteth and warmeth the stomacke, and causeth a good digestion.' Parkinson referred to the choke-pear (the name indicating its unpalatable nature) in 1629, which at that time applied to any wild, very astringent type of pear: 'The Perry made of Choke Pears, notwithstanding the harshness and evill taste, both of the fruit and juice, after a few months, becomes as milde and pleasante as wine.'

During the eighteenth and nineteenth centuries perry was frequently sent down to London and Bristol where it either masqueraded as champagne or was added to inferior, imported wines to improve their quality. The popularity of the drink in Britain has declined since then, except for a while in the 1950s and '60s when a clever advertising campaign encouraged young women to say, 'I'd rather have a Babycham.' Though I am sure no one was fooled into thinking they were ordering real champagne, I wonder how many people realized that they were drinking perry.

Although perry is produced in a similar way to cider, it is traditionally made in single- variety batches, with the blending being done after fermentation. Famous English varieties include Arlingham Squash, Taynton Squash and Barland, whilst the names of others hint at the potency of the brew – Merrylegs, Mumblehead, Lumberskull, Drunkers and Devildrink.

To see a fine old orchard full of towering perry pear trees, with their distinctive arching boughs, heaving with piercing white blossom in the spring or bent heavy with thousands of

An elder in flower. Thought by many to be merely a weed, this is one of our most useful native trees, providing food, drink, medicinal cures and several dyes. Elderflower cordial has achieved unprecedented popularity in recent years, leading to cultivation of the tree.

tiny pears in the autumn, is a marvellous experience. Many such orchards have been lost and few have been replanted but there are enough enthusiasts out there to ensure that they don't die out altogether. Perhaps we can draw our inspiration from an individual living tree that once grew in Holme Lacy near Hereford. It was once a colossal tree dating back to at least 1790 when, it was claimed, it covered about 3⁄4 acre and regularly produced crops of between 5 and 7 tonnes of pears a year. Although recent research has determined that the current tree is not the same specimen recorded in the eighteenth and nineteeth centuries (this was on the edge of the rectory garden and has now disappeared) this tree is remarkably close by and has adopted a very similar mode of survival, currently with nine layered stems, although the original bole may well have perished. However, as a clonal survivor one would have to estimate its age to be at least 200 years and possibly a lot older.

It is fairly certain that the Romans brought dessert pears with them to Britain, though there are few firm records of their cultivation until the late thirteenth century. There are court accounts of the period recording the import of the fruit from France – Eleanor of Provence, wife of Henry III, definitely planted pears of French origin in her garden. France and Belgium have been the principal sources for most of the varieties of pears known in Britain today and this is reflected in their names: Williams' Bon Chrétien and Beurré Hardy, for example. The most famous English-grown variety is probably the Conference. Thomas Rivers raised this at the Sawbridgeworth nurseries in Hertfordshire and gave it its name when he first introduced it to the public at the National British Pear Conference of 1885.

THE ORCHARD LEGACY

Little more than 50 years ago the orchard legacy left for us to tend and pass on to future generations was something still extensive and special. Until recently, as a nation, we seemed to have simply ignored or forgotten this or, at worst, we have eradicated much of it because something else was more important for the land or there was better potential for a financial return. Old-fashioned fruit trees were simply irrelevant in an economy based on the Common Agricultural Policy and dominated by the supermarkets.

Thankfully, there are just enough people left who care about Britain's fruit heritage, the distinctive orchard landscapes, the cultural ties and the wildlife habitats to repair some of the damage and to create new orchard havens. With luck, hard work and a following wind there will always be cob-nuts in Kent, mazzard cherries in the Tamar Valley and plums across the Vale of Evesham.

Some new trends in orcharding are slowly emerging. Elder, considered by many to be a weed, is now valued for its aromatic flowers to make elderflower cordial. Manufacturers used to rely on freelance, seasonal pickers for the flowers but demand has grown so much that one Leicestershire company has planted its own orchard of elder to meet its requirements. With the onset of global warming it will be interesting to see how fruit growing changes in the future and whether peaches, apricots and figs will form part of a new orchard culture.

... AND ON A PERSONAL NOTE

I feel moved to put on record my abiding affection for my family's orchard at our home in Herefordshire. When we viewed a semi-derelict farmhouse with 4 acres of neglected land some 30 years ago, it was the orchard that was one of the clinchers for me. I don't think it had seen a moment's care for more than ten years. Some trees lay windthrown, others had lost boughs, and great balls of mistletoe clogged their crowns. The grass beneath was long and rank and, in its mid-winter dormancy, the whole area might have seemed worthy of little except clearing out and felling. Walking around, though, I began to sense the significance of the place as I felt the presence of the great, gnarled and individual old trees. I found mementoes in the shape of baler-twine graftings fashioned by the old boy who had once owned this land. I watched with wonder as hundreds of hungry fieldfares plundered the autumn's forgotten harvest. After that I had absolutely no hesitation: this was the place.

Since then we have pruned the trees (gradually, so as not to traumatize the older specimens), filling in the gaps with traditional varieties of apple and perry pear grafted on to standard rootstocks. We have introduced grazing sheep to regenerate the nutrients in the soil naturally, and our annual hay harvest allows many species of wild flowers to flourish. There is no measure for the joy that an orchard brings: finding the first cowslips in spring, the day in June when I discovered our very first spotted orchid (from where it came I know not), watching woodpeckers flitting among the trees, and hearing the unearthly screech of the little owl at night. I can lie beneath the sweet-scented, pink- blossomed boughs in May and dream of bringing in the orchard's bounty at the end of the season – Bramleys for pies and crumbles, damsons for wine, pears for perry, and a few boxes of Blenheims see us through the winter.

Gazetteer

SOUTH WEST

DEVON

Bicton Park and Bicton College Garden, near Sidmouth, Devon: www.bictongardens.co.uk or Tel: 01395 568465

Bicton Park has over 60 acres of gardens, including an arboretum. Bicton College Garden & Arboretum is set in Grade I listed parkland approached along a unique avenue of monkey puzzle trees. There is a fine arboretum here, which has a half mile walk, magnificent in spring. Bicton Park Botanical Gardens, East Budleigh, Budleigh Salterton, is nearby.

Orchard Trails

In the springtime the orchards in Devon (*see also* Herefordshire and Worcestershire) burst into flower and are a sight to behold. Many landowners and farmers offer guided tours around the orchards. In autumn return to sample some of the luscious fruit often available for sale at the farm gate or local farm shops.

DORSET

The Martyr Tree, Tolpuddle, Dorset

Britain's most famous sycamore, in the village of Tolpuddle, where a group of agricultural workers (later dubbed the Tolpuddle Martyrs) met to form the very first trade union in 1834 and were deported as a punishment for seeking a living wage. The tree is now in the care of the National Trust (see page 149).

GLOUCESTERSHIRE

Batsford Arboretum, Batsford Park, Moreton-in-Marsh, GL56 9QB. Located 1.25 miles west of Moreton-in-Marsh: www.batsarb.co.uk or Tel: 01386 701441
Located in the picturesque Cotswolds, this collection of trees and plants was begun by Algernon 'Bertie' Mitford in the late nineteenth century. It now boasts the National Collection of Japanese flowering cherries.

Speech House Woodland and Cyril Hart Arboretum, Forest of Dean.
Both a huge oak-dominated forest through which you can ramble, ride or cycle and a formal arboretum.

Tortworth Chestnut, Tortworth
Located near the village church of Tortworth, probably the oldest and largest broadleaf tree in Britain, with a huge and monstrous bole (see page 149).

Westonbirt, The National Arboretum, near Tetbury, GL8 8QS: www.forestry.gov.uk/westonbirt or Tel: 01666 880220
Without doubt the biggest and best arboretum in Britain. More than 3700 different types of trees to enjoy throughout the seasons and especially noted for its national collection of Japanese maples; lots of special events and concerts.

SOUTH AND SOUTH EAST

BUCKINGHAMSHIRE

Burnham Beeches National Nature Reserve, near Slough
Here are many remarkable ancient beech pollards standing in restored wood-pasture, including the famous Cage Pollard, which featured in the film *Robin Hood, Prince of Thieves* (see page 63).

HAMPSHIRE

The New Forest This is the largest wooded forest in England and now a National Park. A rich tapestry of habitats: heathland, grass plains, conifer stands as well as mature oak and beech woodland.

HERTFORDSHIRE

Ashridge Estate, Visitor Centre, Moneybury Hill, Ringshall, Berkhamsted, HP4 1LX: www.nationaltrust.org.uk or Tel: 01494 755557 (Infoline)

On the Hertfordshire/Buckinghamshire borders, this is a vast area of open downland and woods in a country park along the ridge of the Chiltern Hills. Great mix of native species including old oaks on Aldbury Common, many huge old coppiced and pollarded beeches, great bluebell woods, an old yew avenue and many orchids and fungi to be discovered. Over 50 species of birds breed here plus you can often see red kites.

KENT

Bedgebury, The National Pinetum, Park Lane, Goudhurst, TN17 2SL: www.bedgeburypinetum.org.uk or Tel: 01580 211781

The largest and most complete collection of temperate conifers on one site anywhere in the world. Many national record specimens plus almost 150 rare and endangered species.

The Blean, around Canterbury, Kent: www.theblean.co.uk

These are a wonderful and varied network of ancient woodlands, many of which have been continuously managed for hundreds of years. Noted for the fine and ancient sweet chestnut coppice woods. Great flora and wildlife too.

SURREY

Crowhurst Yew, St George's Church, Crowhurst

This huge ancient yew tree, with a door to the hollow trunk, stands in the church grounds and is reputed to be 4000 years old. It is one of The Tree Council's Fifty Great British Trees (see page 80).

Windsor Great Park, The Windsor Estate

Some 5000 acres of glorious Surrey and Berkshire countryside, once a vast Norman hunting chase. Deer lawns, woodlands, coverts and a marvellous collection of huge old veteran oaks make this one of the most special places on London's doorstep to visit and relax.

SUSSEX

Brighton, especially Preston Park, and Hove
All over Brighton and Hove you can go on an elm voyage of discovery. Streets, private gardens and parks – particularly Preston Park, situated just north of Brighton's city centre – offer a fine array of species. The elms are thriving in neighbouring Eastbourne too.

Kingley Vale, near Chichester, West Sussex
Situated in chalk downs north of Chichester and generally regarded as the finest yew wood in Europe. Many ancient yews in excess of 500 years old. Managed by the Nature Conservancy Council.

INNER AND GREATER LONDON

Parks and Squares There are many throughout the capital. Park highlights include superb weeping willows around the Serpentine in Hyde Park, the Elfin Oak in Kensington Gardens, also Hyde Park, and a splendid tree collection in Battersea Park, just a couple of miles south of Marble Arch. Square highlights include Berkeley Square's London planes, some of which date back to 1789. Also noteworthy are Holland Park, Kensington, Cannizaro Park, Merton, and the City of Westminster Cemetery at Hanwell, Middlesex.

Bushy Park, near Hampton Court Palace, Hampton, Surrey
Located just north of Hampton Court Palace, this second largest of the royal parks is famous for its majestic avenue of horse chestnuts, which, in mid-May, when at their flowering peak, attract visitors for Chestnut Sunday.

Epping Forest, northeast London
The largest open public space in London, where the northeast fringe of the city meets Essex. Great tract of ancient woodland saved for the public by Act of Parliament in 1878. Most noted for its beautiful old beeches and hornbeams but many good oaks too. Management policy of reviving and creating pollards is ongoing.

Royal Botanic Gardens, Kew, Richmond, Surrey TW9 3AB: www.rbgkew.org.uk or Tel: 020 8332 5655
One of the finest botanic gardens in the world; since the eighteenth century Kew has been an evolving combination of landscapes, plant collections, arboretum and historic buildings. Wonderful throughout the year and, since 2003, a UNESCO World Heritage Site.

EAST

CAMBRIDGESHIRE

Cambridge University Botanic Garden, Cory Lodge, Bateman Street, Cambridge CB2 1JF: www.botanic.cam.ac.uk or Tel: 01223 336265
Less than one mile south of the city centre, some 40 acres of superb gardens offering year-round interest. More than 10,000 labelled plant species (including nine national collections) plus a great tree collection. Particularly noted for its horse chestnuts, native sorbus collection, limes from all over the world (a lime leaf is the garden's logo), majestic cedars, some of Britain's oldest giant redwoods and the first dawn redwood to be grown in Britain, planted in 1949 and deemed one of Fifty Great British Trees in honour of the Queen's Golden Jubilee.

EAST ANGLIA

The Breckland, Norfolk
The Breckland deal rows are to be found along many of the roads to the south and west of Swaffham. These are these unique, abandoned 'hedges' of Scots pines, some planted almost 200 years ago. Remarkable bent and wind-formed shapes (see page 120).

Thetford Forest Park, near Thetford and Brandon
Also in the heart of Breckland, on the Norfolk-Suffolk borders, this is a patchwork of pines, heathland and broadleaves and Britain's largest lowland pine forest.

ESSEX

Hatfield Forest, Takeley, near Bishop's Stortford, Essex CM22 6NE: www.nationaltrust.org.uk or Tel: 01279 874040 (Infoline)
Some 1049 acres, one of the finest examples of a medieval royal hunting forest comprising ancient grazed woodpasture and coppiced woodland, continuously managed as such for over 1000 years. Many fine native trees, but most especially noted for great array of fine old coppiced and pollarded hornbeams. (Future expansion of Stansted Airport currently threatens the many rare and delicate habitats in this special place, a designated Site of Special Scientific Interest and a National Nature Reserve.)

EAST AND WEST MIDLANDS

HEREFORDSHIRE

Croft Castle, Yarpole, near Leominster, HR6 9PW: www.nationaltrust.org.uk or Tel: 01568 780141 (Infoline)
Famous for its huge sweet chestnuts (some of which form Britain's only triple avenue of these trees, the Armada Avenue) and oaks, many of which are about 450 years old (see pages 149-51).

Much Marcle Yew, Hellens and St Bartholomew's Church, Much Marcle
One of the oldest and biggest yews in Britain – famously has a seat within where several people can sit (see page 80).

Orchard Trails
In the springtime the orchards in Herefordshire and Worcestershire, especially around Evesham (*see also* Devon), burst into flower and are truly a sight to behold. Many landowners and farmers offer guided tours around the orchards. In autumn return to sample some of the luscious fruit often available for sale at the farm gate or local farm shops.

LINCOLNSHIRE

Bowthorpe Oak, Manthorpe, Bourne
Located at Bowthorpe Park Farm, three miles southwest of Bourne, this hollow tree is notable as the very biggest English oak in Europe with a stupendous girth of 42 feet (see page 29).

NOTTINGHAMSHIRE

Clumber Park, Worksop S80 3AZ: www.nationaltrust.org.uk or Tel: 01909 544917 (Infoline)

Just north of Sherwood Forest, the estate boasts an almost 2-mile-long avenue that consists of 1296 lime trees (Tilia x europaea), planted in about 1840, in a double row on each side of the drive. It's the longest avenue of common limes in Europe.

Sherwood Forest

Edwinstowe, the central village of the Forest, composed of several woods, is a good place to start to explore what was supposedly once the haunt of Robin Hood & co. The star attraction is the ancient Major Oak, probably well in excess of 500 years old. It is to be found in the Birklands, the wood that contains the remnants of the once extensive oak forest that used to cover most of Sherwood Forest. In the surrounding woodland are many other similar (though not quite as large) oaks with much birch in attendance.

WORCESTERSHIRE

Orchard Trails, *see* Herefordshire

NORTH WEST

CUMBRIA

Levens Hall and Gardens, Kendal, Cumbria LA8 OPD: www.levenshall.co.uk or Tel: 015395 60321

Located five miles south of Kendal, this is without doubt Britain's finest topiary garden, dating back to 1694 (see page 88).

Borrowdale Woods, south of Keswick and alongside Derwentwater, Cumbria.

A whole network of subtly different woodlands well worth exploring, principally for the 'Atlantic' oak woods, rich in mosses and lichens but also for many other native species as well as impressive stands of introduced conifers.

SCOTLAND

ARGYLL

Benmore Botanic Garden, Dunoon, Argyll PA23 8QU: www.rbge.org.uk or Tel: 01369 706261

Given to Royal Botanic Garden, Edinburgh, in 1929 the gardens are 7 miles north of Dunoon on the Cowal Peninsula, overlooking the Holy Loch and the Eachaig valley. The gardens were an ideal west-coast satellite and are noted for their splendid rhododendron displays, a mighty Sierra redwood avenue and the Chilean rainforest glade, home to all manner of exotic conifers and southern beeches.

MIDLOTHIAN

Royal Botanic Garden, Edinburgh: www.rbge.org.uk or Tel: 0131 552 7171

One of Scotland's premier collections of trees and plants. Founded in the seventeenth century and covering 31 acres, this is a world-renowned scientific centre for the study of plants, their diversity and conservation.

PERTHSHIRE

Big Tree Country, Perthshire

This region of Scotland boasts some of the biggest and best of the introduced conifers, including larch, Douglas fir and silver fir. See also the Black Wood of Rannoch, the most extensive area of relict Caledonian forest remaining in Perthshire.

The Fortingall Yew, near Aberfeldy, Perthshire

Generally considered the oldest tree in Britain, if not Europe. Quite possibly the tree is about 5000 years old, but only a fragment of the original bole remains (see page 77).

Scone Palace, Perth PH2 6BD: www.scone-palace.co.uk or Tel: 01738 552300

Located just north of Perth, the forests and woodland of Scone Palace boast some of the very first Douglas firs planted in Britain in 1834, some huge Sitka spruces, many other conifers and a massive sycamore reputedly planted by King James VI of Scotland and I of England in 1617.

Caledonian Pine Forests

These are the remnants of what was reputedly a vast tract of native pinewoods across the Highlands. Sites to visit include Rothiemurchus and Abernethy on Speyside, Glen Affric to the southwest of Inverness, Glen Tanar on Deeside, near Aboyne, Ballochbuie Forest, near Braemar, and the Cairngorm Glens in Mar.

WALES

CARMARTHENSHIRE

Bute Park Arboretum, Cardiff

Once part of the grounds of Cardiff Castle, this wonderful, urban, green space, in the heart of Cardiff city centre, has one of the finest collections of trees in any British municipal park. Many of the trees are known to be the biggest examples of their species anywhere in the UK. An interesting mix of rare and ornamental trees have been planted since the 1940s, which compliment some excellent specimens that formed part of the original park design. Free access to all.

Talley Abbey Ash, Llandeilo

By the remains of the twelfth-century Talley Abbey, north of Llandeilo, one of the very biggest ash trees in Britain with a girth of 36 feet (see page 46).

CEREDIGION

Hafod Estate, Pontrhydygroes, Ystrad-Meurig SY25 6DX: www.hafod.org or Tel: 01974 282568

In the Yswyth Valley, near Devil's Bridge, just 12 miles southeast of Aberystwyth, this is superb example of Picturesque landscape, created by its owner Thomas Johnes in the late eighteenth century, covers some 200 hectares. Stupendous woodland walks with some massive beeches to seek out. A restoration programme is in progress under the partnership of the Hafod Trust and Forestry Commission Wales.

NORTHERN IRELAND

COUNTY FERMANAGH

Crom Yews, Crom Estate, Upper Lough Erne, Newtownbutler, BT92 8AP: www.nationaltrust.org.uk or Tel: 028 6773 8118 (Visitor centre)
In one of Ireland's most important nature conservation areas, the branches of two massive yew trees (a male and a female) interweave to form one gigantic green canopy, below which the lower boughs gyrate and snake this way and that on the floor beneath (see page 16).

Original Irish Yew in Florence Court, Enniskillen, BT92 1DB: wwww.nationaltrust.org or Tel: 028 6634 8249
Located south of Enniskillen, in Florence Court Forest Park, this is the mother tree of all Irish yews (distinguished by the upthrust form of the boughs), common throughout parks and churchyards nationwide (see page 90).

Bibliography

Ablett, William H., *English Trees and Tree-planting* (Smith, Elder & Co., 1880)

Barnes, Gerry, and Williamson, Tom, *Hedgerow History* (Windgather Press, 2006)

Bevan-Jones, Robert, *The Ancient Yew* (Windgather Press, 2002)

Bingham, Madeleine, *The Making of Kew* (Michael Joseph, 1975)

Boulger, G.S., *Familiar Trees* (Cassell & Co., 1886)

Clark, Ethne, and Wright, George, *English Topiary Gardens* (Weidenfeld & Nicolson, 1988)

Clouston, Brian, and Stansfield, Kathy, *After the Elm* (Heinemann with The Tree Council, 1979)

Cobbett, William *Rural Rides* (Cambridge University Press, 1922)

Common Ground, *Local Distinctiveness* (Common Ground, 1993)

Common Ground, *Orchards – A Guide to Local Conservation* (Common Ground, 1989)

Common Ground, *the common ground book of Orchards – conservation, culture and community* (Common Ground, 2000)

Cresswell, Ruth Alston, *Spirit of the Trees* (Society of The Men of the Trees, 1947)

Culpeper, Nicholas, *Complete Herbal* (London, 1653; several subsequent editions including Wordsworth Editions Ltd., 1995)

Edlin, H.L., *British Woodland Trees* (B.T. Batsford Ltd., 1944)

Elwes, H.J. and Henry, A., *The Trees of Great Britain and Ireland* (Edinburgh, 1908-13)

Evelyn, John, *Silva – or A Discourse of Forest-trees* (1st Hunter edition, York, 1776)

Flanagan, Mark, and Kirkham, Tony, *Plants from the Edge of the World* (Timber Press, 2005)

Gerard, John, *The Herball, on Generall Historie of Plantes* (John Norton, 1597)

Grigson, Geoffrey, *The Englishman's Flora* (Helicon Publishing, 1996)

Hadfield, Miles, *British Trees* (J.M. Dent & Sons, 1957)

Harris, Esmond, Harris, Jeanette, and James, N.D.G., *Oak – A British History* (Windgather Press, 2003)

Hudson, W.H., *Nature in Downland* (Dent, 1923)

James, N.D.G., *A Book of Trees* (The Royal Forestry Society, 1973)

Johns, The Revd C.A., *The Forest Trees of Britain* (SPCK, 1882)

Johnson, Hugh, *The International Book of Trees* (Mitchell Beazley, 1973)

Jordan, Michael, *The Green Mantle* (Cassell & Co., 2001)

Knight, Thomas Andrew, *Pomona Herefordiensis* (Birdsall & Son, 1811)

Lewington, Anna, and Parker, Edward, *Ancient Trees* (Collins & Brown, 1999)

Linnard, William, *Welsh Woods and Forests* (Gomer Press, 2000)

Loudon, J.C., *Arboretum et Fruticetum Britannicum or The Trees & Shrubs of Britain* (London, 1838)

Luckwill, L.C. and Pollard, A., *Perry Pears* (published by the University of Bristol for the National Fruit and Cider Institute, 1963)

Mabey, Richard, *Flora Britannica* (Sinclair-Stevenson, 1996)

Marren, Peter, *The Wild Woods* (David & Charles, 1992)

Miles, Archie, *Silva* (Ebury Press, 1999)

Milner, J. Edward, *The Tree Book* (Collins & Brown, 1992)

More, David, and White, John, *The Illustrated Encyclopedia of Trees* (Timber Press, 2002)

Morton, Andrew, *The Trees of Shropshire* (Airlife, 1986)

Muir, Richard, and Muir, Nina, *Hedgerows – Their History and Wildlife* (Michael Joseph, 1987)

Musgrave, T., Gardner, C., and Musgrave, W., *The Plant Hunters* (Ward Lock, 1998)

Pakenham, Thomas, *Meetings with Remarkable Trees* (Weidenfeld & Nicolson, 1996)

Pakenham, Thomas, *Remarkable Trees of the World* (Weidenfeld & Nicolson, 2002)

Pennant, Thomas, *A Tour of Scotland in 1769* (Melven Press, 1979. First edition printed in Warrington by W. Eyres, 1774)

The Penny Magazine, various articles & images (Charles Knight 1832, 33, 35, 42 & 43). Copies scarce, but available in old bookshops.

Philpott, Hugh B., *Britain at Work* (Cassell & Co., circa 1900)

Pollard, E., Hooper, M.D., and Moore, N.W., *Hedges* (Collins, 1974)

Rackham, Oliver, *The Illustrated History of the Countryside* (Seven Dials/ Cassell & Co., 2002)

Rodgers, John, *The English Woodland* (B.T. Batsford & Co., 1941)

Saturday Magazine, The, various articles and images (John William Parker, 1836-7). Copies scarce, but available in old bookshops.

Selby, Prideaux John, *A History of British Forest Trees* (John Van Voorst, 1842)

Stokes, Jon, and Hand, Kevin, *The Hedge Tree Handbook* (The Tree Council, 2004)

Stokes, J., Rogers, D., Miles, A. and Parker, E., *The Heritage Trees of Britain and Northern Ireland* (Constable with The Tree Council, 2004)

Stokes, J., White, J., Miles, A. and Patch, D., *Trees in Your Ground* (The Tree Council, 2005)

Strutt, Jacob George, *Sylva Britannica* (Longman, Rees, Orme, Brown & Green, 1830)

White, The Revd Gilbert, *The Natural History of Selborne* (Penguin Classics, 1977)

Whitlock, Ralph, *A Calendar of Country Customs* (B.T. Batsford Ltd., 1978)

Wilkinson, Gerald, *Epitaph for the Elm* (Book Club Associates, 1978)

Wilkinson, Gerald, *A History of Britain's Trees* (Hutchinson & Co. Ltd., 1981)

Williamson, Tom, *Hedges and Walls* (The National Trust, 2002)

Woodward, Marcus, *The New Book of Trees* (A.M. Philpot, circa 1920)